# LINKING NETWORKS: THE FORMATION OF COMMON STANDARDS AND VISIONS FOR INFRASTRUCTURE DEVELOPMENT

# Linking Networks:
# The Formation of Common Standards and Visions for Infrastructure Development

*Edited by*

MARTIN SCHIEFELBUSCH AND HANS-LIUDGER DIENEL
*Berlin Technical University, Germany*

Routledge
Taylor & Francis Group

LONDON AND NEW YORK

First published 2014 by Ashgate Publishing

2 Park Square, Milton Park, Abingdon, Oxfordshire OX14 4RN
711 Third Avenue, New York, NY 10017

*Routledge is an imprint of the Taylor & Francis Group, an informa business*

First issued in paperback 2018

**British Library Cataloguing in Publication Data**
A catalogue record for this book is available from the British Library.

**The Library of Congress has cataloged the printed edition as follows:**
Linking networks : the formation of common standards and visions for infrastructure development / [edited] by Martin Schiefelbusch and Hans-Liudger Dienel.
    pages cm -- (Transport and society) Includes bibliographical references and index.
 ISBN 978-1-4094-3920-2 (hardback) 1. Infrastructure (Economics)--Europe.
2. Infrastructure (Economics)--Europe--History--19th century. 3. Infrastructure (Economics)--Europe--History--20th century. 4. European cooperation. I. Schiefelbusch, Martin, editor of compilation. II. Dienel, Hans-Liudger, 1961- editor of compilation.
III. Series: Transport and society.
 HC240.9.C3L56 2014
 338.94--dc23

                                                                   2014010309

ISBN 978-1-4094-3920-2 (hbk)
ISBN 978-1-138-54652-3 (pbk)

# Contents

**PART III        COMMENTS**

**PART IV        CONCLUSIONS**

# List of Figures

# List of Tables

# Notes on Contributors

### Editors

**Martin Schiefelbusch** is researcher in transport and mobility and Area Manager at the nexus Institute in Berlin and the Centre for Technology and Society of Berlin Technical University. In addition to the topics covered by this volume, his research interests are transport policy, planning theory, public transport development and the representation of citizens' interests in transport planning and policy. He studied transport planning and geography in Berlin and London. His PhD, completed in 2007, dealt with the incorporation of the travel experience in transport planning processes and public transport services. He has published extensively on this topic as well as on leisure travel, participation, customer care and transport policy issues.

**Hans-Liudger Dienel** is Professor for Work, Technology and Society at Berlin University of Technology and director of the nexus Institute in Berlin. Before this he was director of the Center for Technology and Society at the same university. After his degrees in history, sociology, philosophy and mechanical engineering (1988/1990), he completed his PhD at the Technical University Munich in 1993. Dienel is a prolific writer on the history and future of transport and infrastructures and is currently president of the International Association for the History of Transport, Travel and Mobility (www.t2m.org).

### Contributors

**Gerold Ambrosius** was professor of Economic History at the University of Konstanz before joining the faculty at the University of Siegen in 2000. He received his doctoral degree from the University of Tübingen and his Habilitation from the Free University of Berlin. In Siegen he teaches Economic and Social History from the seventeenth to the twentieth century. He is inter alia member of the advisory board of the Bundesverband öffentlicher Dienstleistungen (Association of public service providers). His main research areas are European economic history, the history of public enterprises including the regulation of public services, the history of institutional competition and the history of economic policy.

**Irene Anastasiadou** was awarded her PhD in 2009 from the Eindhoven University of Technology for her thesis 'In Search of a Railway Europe; Transnational Railway Developments in Interwar Europe'. Since then she has undertaken research in the context of various research projects. At the moment she is working as a postdoctoral researcher at the Faculty of Technology, Policy and Management at the Delft University of Technology (POLG). Her research focuses on the governance of the natural gas sector in Europe from a historical and transnational perspective. From September 2014 she will work as an IPODI Marie Curie Fellow at the Technical University of Berlin on the topic of 'Railways and Europe - Asian Relations, 1940s–Present'.

**Jens Ivo Engels**, studied Modern and Contemporary History at the Universities of Freiburg and Bordeaux. His positions include assistant professor 1998–2004 and associate professor 2005–2008 at Freiburg. Since 2008 he has been professor of Modern and Contemporary History at Technische Universität Darmstadt. His research interests include the history of infrastructure, corruption history, environmental history, the history of social movements and the history of early modern monarchy.

**Klaus Gestwa** is professor and director of the Institut for Eastern European History and Area Studies at the University of Tübingen. His Habilitation thesis on Stalin's mega projects makes an important contribution to Soviet history after 1945. His research also comprises the cultural history of the Cold War, the history of the Eastern Bloc and the global history of technology and science. He is currently working on monographs on the history of the Soviet Gulag and on Great Constructions Sites in Eastern Europe since 1945.

**Melissa Gómez** is mobility and service innovation researcher at Citymart and previously worked as researcher in mobility and demography at the nexus Institute in Berlin. She studied Political Science in the Complutense University of Madrid and Urban Planning at the Barcelona University and the Institute of Urbanism of Paris. Her research interests are transport planning and mobility strategies and urban governance and participation. She has worked in several organisations who promote sustainable means of transportation in cities such London and Barcelona and has carried out several projects in urban accessibility for the Catalan government and other Spanish administrations.

**Christian Henrich-Franke** works at the Department for History and the 'Institut für Europäische Regionalforschungen' at the University of Siegen. His main fields of research are International Relations (European Integration), Radio/Broadcasting History and Governance. His PhD dealt with the global regulation of radio frequencies between 1945 and 1988. His Habilitation compared transport policy in the German Reich (1871–1878) with transport policy in the European Economic Community (1958–1972).

**Jacqueline Klopp** is Associate Research Scholar at the Center for Sustainable Urban Development and Adjunct Professor at the School of International and Public Affairs of Columbia University, New York. She received a BA at Harvard

University and obtained a PhD in Comparative Politics and Political Theory from McGill University in 2001. She has published widely on international and African politics and is currently involved in the Volvo Foundation's center for transport research in developing countries.

**Tomasz Komornicki** is associate professor and deputy Director at the Institute of Geography and Spatial Organisation at the Polish Academy of Sciences (IGSO PAS). He studied Geography in Warsaw. His main topics of interest are transport geography, transport policy, spatial planning and development of border regions. He published several books and participated in a number of Polish and European research projects (including six ESPON Projects). Furthermore, he is member of the Consulting Board in the Ministry of Regional Development for preparation of the new Concept of Spatial Development of Poland up to 2030.

**Vincent Lagendijk** studied economic history at Leiden University, and wrote a PhD dissertation on the history of the European electricity network at Eindhoven University of Technology. He currently works at Maastricht University, writing a monograph on the transnational history of the Tennessee Valley Authority.

**George Makujama** is a transport engineer working with the African Development Bank in Nairobi. He has spent over 12 years as a practicing engineer in Africa and as a researcher with the Volvo Foundation funded Urban Transport Studies in Africa with the ACET centre based at Cape Town University. Makajuma holds an MSc in Urban Engineering with a specialisation in Urban Transport and MBA degree.

**Matthias Ruete** is currently Director General in the Directorate General for Mobility and Transport of the European Commission. Before this he held various senior management and cabinet positions in the Commission covering a wide range of issues, such as enlargement, transport, internal market, industrial policy and research. Before joining the European Commission, he lectured in Constitutional, European and International Public Law at Warwick University.

**Bruce E. Seely** is a historian of technology whose research interests include the history of transportation and transportation policy. He has authored numerous published on the development of the highway system in the United States, including 'The Best Transportation System in the World: Railroads, Trucks, Airlines, and American Public Policy in the Twentieth Century', with co-authors Mark Rose and Paul Barrett (2006). He has taught at Michigan Technological University since 1986 and is currently Dean of the College of Sciences and Arts.

**Aristotle Tympas** teaches undergraduate and graduate courses in 'History, Technology, Society' and 'Science, Technology, Society' at the National and Kapodistrian University of Athens Greece. He specialises in the study of the digitalisation-society and the technology-environment relationships.

**S. Waqar Zaidi** is Assistant Professor in European History at the Lahore University of Management Sciences (Pakistan). Prior to this he was visiting researcher at Imperial College London. He completed his PhD at the Centre for the History of Science, Technology, and Medicine at Imperial College London, on 'Proposals for the Internationalisation of Aviation and the International Control of Atomic Energy in Britain, USA and France'. He is currently working on a monograph on technology and internationalism.

Foreword

# Integration of Infrastructures –
# Comparative Analyses

Matthias Ruete[1]

In Europe, there is a tradition of more than 2,000 years of building transport networks. Roman roads, as the backbone of the Roman Empire, determined the location of market places and cities, most of which still exist. Today, the promotion of a trans-European transport network (TEN-T), introduced by the Maastricht Treaty in 1992, has played a key role for European integration and establishing the common market, starting with the adoption of the first TEN-T Guidelines in 1996.

An in-depth review of the TEN-T policy is now under way. In October 2011, the Commission presented its proposal for a new TEN-T Regulation, which is now passing through the decision-making process and might come into force in autumn 2013 – together with the 'Connecting Europe Facility' (CEF), which will be the corresponding funding and financing instrument.

With these proposals, the Commission puts forward for the first time the idea of a transport network for the EU as a whole, designed on the basis of a uniform methodology with a view to responding to European needs. It takes into account, in a balanced way, cohesion aspects and traffic flows, the economy and environment. All modes of transport are covered and the most advanced technologies will be deployed for traffic management and de-carbonisation. The goal is to go beyond national borders, creating a single European transport system as the backbone of the internal market.

Based on a dense 'comprehensive' network identified by Member States, which comprises a number of sea and inland ports, road–rail terminals and airports, the Commission has selected those parts of this network which have the highest value for Europe, to shape the 'core' network.

The core network includes the most important rail and road links but also other transport modes such as inland waterways with transnational importance, and the most important airports and ports. Both comprehensive and core networks will be supplemented by Motorways of the Sea, which represent the maritime dimension of the TEN-T, to connect insular Member States, other important islands and peninsulas.

---

1    Mr Ruete is currently Director General in the Directorate General for Mobility and Transport of the European Commission.

Since large parts of the core network already exist, the objective is to construct the remaining parts and upgrade the existing ones, where necessary to meet the required standards, by 2030. The rest of the comprehensive network should be built by 2050.

To achieve this ambitious goal, the CEF Regulation envisages grouping together those parts of the core network where the greatest need for coordination may be expected, into 'corridors' that comprise several modes. Continuing the successful example of current TEN-T priority projects, European Coordinators will be appointed to steer the development of these corridors at a political level.

With this background in mind, the Commission pays particular tribute to the present book, which from a scientific viewpoint sums up and compares the development of infrastructure networks in several parts of the world in different periods, with a focus on issues of international coordination and cooperation.

# PART I
## Introduction and
## Theoretical Background

# Chapter 1
# Introduction and Overview

Martin Schiefelbusch and Hans-Liudger Dienel[1]

Infrastructure has played a key role throughout human history, one of an essentially strategic character in the development of international order and the global economy. The study of its integration is of great interest in understanding the processes of collaboration and exchange between regions or countries, for example its influence on the evolution of international trade and development policies. We could not understand the network of communications, product technology and the development of states, which are intertwined throughout various global connections platforms, without understanding the development of trade and economic networks. Behind this structure lies an assembly of social, cultural and political history which has, to a certain extent, permitted a tacit cohesion of infrastructure networks. This process has, in the case of Europe, opened the door to the infrastructure networks we have today.

In the globalised economy, the productivity of distinctive factors upon the efficiency of markets and countries trade services, both internal and external, are a function of many variables. With this in mind, a strong multidisciplinary and comparative analytical approach to solutions for comprehensive development is required. In this regard, the proposal raised by this book is to analyse how the relationships between different (usually national) infrastructure networks can be designed so that networks are created as a holistic global system and not as disjointed parts.

During the second half of the nineteenth century and the middle of the twentieth, the global scene was dominated by two fundamental phenomena: unstoppable technological progress and the ascension of the global economy. Markets tended to develop considerably faster than political and national structures, to the point that a supranational system was imposed to the detriment of nation states, promoting significantly more important territorial interdependence beyond the geographically delimited borders. It is therefore important to understand how national entities went or did not go from wanting to protect and favour their local markets, to seeing an opportunity to boost their own chances through international integration.

If we look at the fundamental changes that affected and influenced the global economy, such as the two world wars and the oil crisis, we see that increasing

---

1 The authors gratefully acknowledge the assistance of Melissa Gómez in the preparation of this introduction.

market competition and the increased trend towards privatisation created by this phenomenon made it necessary to generate new and integrative concepts in order to respond to those challenges. For this reason, the study of infrastructure integration can be regarded as one of the most important elements in understanding a political and economical situation with worldwide distribution chains, as well as the insertion and positioning of the diverse actors therein.

This book is part of an emerging research area dealing with the mechanisms of international collaboration. It strengthens comparative research as a complement to the detailed analysis of singular cases that often characterises previous works in this field. This wide range of topics will enable the reader to get a good overview of the different challenges posed and the strategies employed in each sector to establish internationally compatible networks, procedures and standards. Yet the main aim of the book is to promote the understanding of cross-sectoral and cross-period comparisons. The examples provided come from different sectors (for example road and rail networks, post and communications and pipelines) and cover the period from the mid nineteenth to late twentieth century. Considering the growing intensity of international collaboration and exchange in many parts of social and economic life, this subject is also of topical interest.

This book includes a series of case studies about the evolution of international infrastructure networks through comparative analyses that have been conducted in various parts of the world and in different periods of time, as well as more general theoretical background articles. The reader will understand the importance of infrastructure integration processes on a global stage, as well strategies and instruments commonly used to carry them out. The challenges that need to be overcome in every stage of the process will be explored, as well as which factors promote and hinder progress.

In this work, the compilation of articles aims to show, among other things, the importance that connectivity has had in facilitating trade, the strategic character that the regulatory policies have had in this regard, and the pre-eminence of the economic dimension – over the political, social and cultural ones – in the various processes of integration. Furthermore, the significance of standards as a key element and their influence on integration processes will be illustrated as well as the integration of these infrastructures as coordinated actions to complement national projects. Besides that, the studies will examine how that may have helped to increase or decrease territorial disparities.

The structure of the book is elaborated in four groups of articles: First, the introduction intents to be an overture to the diverse theoretical concepts. The main part of the book consists of case studies, presented in comparative analyses on different infrastructure sectors such as road and rail networks, electricity and telecommunication. As much of the empirical work is based on Western Europe, the third section aims to enrich the case studies by giving additional perspectives from other countries' experiences. The fourth and last part summarises and discusses from an analytical approach the results found and presents the future prospects for the research and the implications for contemporary policy development.

In the introductory and theoretical background sections of this text, the reader will find articles which explore the methodological and conceptual challenges of comparisons across sectors and periods of the integration of infrastructure using a pragmatic approach by Christian Henrich-Franke. Critical reflections about the integration paradigm which shows its effects on a local, national and transnational level will be provided by Jens Ivo Engels, who also discusses the issues of fragmentation and segregation. Gerold Ambrosius attempts to show whether the archetypes of infrastructural integration between countries exist or not, using a historical-empirical approach and endeavours to identify them.

The case studies' section covers diverse examples of cross-sectoral and cross-period comparisons. On a European stage, Christian Henrich-Franke contributes a study of railway tariff integration, using cases from the German Reich and the European Economic Community. S. Waqar Zaidi shows contested internationalist visions for the governance of post-war European Aviation in Britain in World War II and the early post-war period, considering the role of nation states and private capital. In addition, Vincent Lagendijk, studies the case of international electricity governance in the interwar and post-World War II period, how the integration process due to the change from a liberal to a liberalised system keep it relatively hidden and how the material and institutional dimensions lead to more integrated systems across borders. Another chapter by Christian Henrich-Franke discusses if and how experiences and models of integration were transferred between different areas – the so-called 'spill-over theory'. Martin Schiefelbusch's chapter on visions for the development of international rail services describes how stakeholders in Europe perceived the opportunities offered by 'going international', but also the concerns that were expressed against such a move.

To conclude the section of case studies, the contribution by Tomasz Komornicki deals not with a specific infrastructure sector, but shows the interrelations between transport infrastructure, socio-economic interactions and political boundaries through the example of the borders of contemporary Poland.

Moving beyond the European scale, George Makujama and Jacqueline Klopp study international infrastructure development in Africa, focusing specifically upon how colonial transport projects were constructed in order to respond to colonial strategies of resource extraction as opposed to national priorities. From a cross-period approach, Irene Anastasiadou and Aristotle Tympas compare the attempts of construction of several transnational railway projects, focusing on Europe-Asian relations during the interwar and post-World War II and the role they played in a colonial and post-colonial context.

The two commenting chapters provide additional insights through experiences from outside Europe. Bruce E. Seely examines the experience of infrastructure integration processes in the USA focusing on the transportation system and the challenges it entailed to attempt to link the different transport modes into a seamless network – an idea that is today a common policy objective, but met strong resistance in the USA for a long time. Klaus Gestwa shows how infrastructure played a key role in the capacity development of the Soviet economy. But his contribution also highlights some concealed elements of the soviet civilisation and the role of infrastructures in establishing and maintaining an authoritarian political system.

This compilation of case studies makes this book key to understanding the integration of infrastructure in recent history and the role of infrastructure as a vital element in worldwide economical development. At the same time, the effects and contradictions on the local, regional, national and transnational scale, as well as its role as a political priority and the impact upon contemporary policy that this engenders become apparent. To understand the European Union today as one of the worldwide paradigms of integration is a condition *sine qua non* to allude to the market of goods, capital and workforce, and the part that infrastructure has played on the whole process. Regarding developing countries, the World Bank's Logistics Performance Index maintains that the improvement of infrastructure affects growth and not vice versa – a good logistical performance can add 1 percent to the economic growth rate. With this in mind, it is essential to look how processes have been carried out and in that way, promote or slow down the place that emergent economies have in the global scenario.

Finally, this book may lead readers to consider the current perspectives and future prospects regarding the integration of infrastructure, showing to what extent vitality and evolution depend on the capacity of mutation and adaptation to the new challenges of the constant changing order, as we witnessed from the revitalisation process that lead the railway industry facing competition from car and air. Furthermore, it contributes to comparative historical research tools and shows us what can be learnt and applied in their developing ongoing processes such as the Single European Transport Area who faces important challenges in terms of overcoming the still existent fragmentation in order to face the 50 percent rise of passenger traffic that is estimated by 2050.

## Acknowledgements

The editors would like to thank all those that have contributed to this book, first and foremost the authors who have been eager to share their thoughts with us, cooperative in handling the outcome of the reviewing process and patient in waiting for the final steps to be taken. In addition to those who have been involved as writers, we would like to thank our supporters Sonja Ziener, Melissa Gómez, Neele Reimann-Philipp and Lennart Riechers from the nexus Institute in Berlin for their assistance in translating, proofreading and commenting. We also thank our colleagues from the University of Siegen for the good discussions and productive collaboration in our joint project on the integration of infrastructures in the nineteenth and twentieth century, which provided the background and occasion for work on this book. Last not least, we would like to thank the German Research

Council (DFG) for sponsoring this research, which we hope will be met with interest and a source of inspiration also beyond the German speaking countries by means of the present book.

Chapter 2

# Methodological and Conceptual Challenges of Comparisons across Sectors and Periods

Christian Henrich-Franke

## 1. Introduction

> 'We can't help making historical comparisons, yet we do not have a logical, accepted way to make them. Facile ones abound, partly because most of us are intellectually a bit lazy, partly because (even if spurious) they can help convince one's listeners or readers ...'

With these words the US environmental historian John McNeill (McNeill 2011) commented on historical comparisons in a review of two new volumes by Jared Diamond and James Robinson (Diamond/Robinson 2010) and by Vaclav Smil (Smil 2010). Both deal with the methodological fundaments of the comparative approach to history. Diamond and Robinson's book in particular is a very inspiring contribution to the discussion on comparative history. The authors are trying to coax historians into using the comparative approach to history, although its scientific value is still a point of much controversy among historians. This controversy merits a separate discussion, but it is beyond the scope of this chapter, and other publications have dealt with it extensively (Kaelble 1999, 27–35).

This chapter focuses on some practical aspects of historical comparisons across sectors and periods. It is not an in-depth reflection on methodological fundaments or typologies. Accordingly, in a first step, the chapter will address the methodological challenges of comparisons across sectors and periods before discussing the conceptual challenges in more detail using a pragmatic approach.

It should be noted that the boundaries between the term 'historical comparison' and the term 'comparison' used in the social sciences are often blurred, in contemporary history in particular. However, in contrast to the social sciences, a suitable reference to the context is usually an integral part of a historical comparison. In most cases historians hope to achieve a complex explanation of a historical phenomenon or development, avoiding the deliberate reduction to a limited number of variables which is at the core of the comparative approach in the social sciences. 'Monocausality is a swearword among historians' to quote Hartmut Kaelble on the subject (Kaelble 1999, 113).

## 2. Methodological Challenges of Comparisons across Sectors and Periods

*2.1 Aims of Historical Comparisons*

No comparison should just be made for its own sake. Therefore each comparative study of a historical topic requires a clearly defined aim. What is it we want to find out by making a comparison? According to Jürgen Kocka, comparisons 'discuss two or more historical phenomena systematically with respect to their similarities and differences in order to reach certain intellectual aims' (Kocka 2003, 39). Consequently, the purpose of a comparison depends on a specific research interest, and the comparison is just a means to achieve an aim.

A large number of different aims has been discussed which are beyond the scope of this chapter (see Haupt/Kocka 1996). Kaelble for example makes a very basic distinction between generalising and individualising comparisons. These are not mutually exclusive, quite the contrary: generalising comparisons hope to identify causations or general rules in historical developments in order to establish a theory or typology, and yet differences have to be taken into account. By contrast, individualising comparisons are used to carve out differences in historical phenomena or developments. In this case there is no intention of establishing general rules or theories, but considering similarities is a precondition because comparing 'apples and oranges' would make no sense (Kaelble 1999, 27).

*2.2 Functions of Historical Comparisons*

Jürgen Kocka describes four different functions of comparisons (Kocka 2003, 40–41):

*2.2.1 Heuristic*
The heuristic function of comparisons enables us to identify questions and issues we might have missed, neglected or ignored if we had just focused on a single event, phenomenon or historical development. In other words, comparisons reveal alternative ways of approaching sources.

*2.2.2 Descriptive*
Viewed from the narrative of history, comparisons help sharpen the profile of historical cases or phenomena. By contrasting different cases, distinct features and similarities become more visible.

*2.2.3 Paradigmatic*
Comparisons provide us with new perspectives not only in our approach to sources, but also on historical cases and developments. This is partly due to the fact that comparisons often cross boundaries between research communities and disciplines. Viewed from an alternative angle, well-established matters often appear in a very different light, and traditional explanations may seem

questionable. Comparisons are vital to broadening the horizon of communities across national borders or disciplines.

### 2.2.4 Analytical

Comparisons are essential in order to establish historical causations, and they allow us to detect causes for historical phenomena or developments and to point out conflicts. They could be seen as 'indirect experiments' used to facilitate testing hypotheses or theoretical models. Frequently based on standardised 'parameters of comparison', comparisons are still quite unpopular among those historians who emphasise the complexity of history. Their advantage is the limitation to a small number of cases. Historical comparisons 'offer insight and opportunities for stronger statements about causations than historians ... usually provide' (McNeill 2010). The strongest argument for using comparisons is the possibility to isolate causations and to establish variations, typologies and theories.

### 2.3 Characteristics of Comparisons across Sectors and Periods

If we accept that historical comparisons fulfil a variety of valuable functions in historical research, in the next step we should establish their characteristics. Jürgen Kocka identifies three overriding characteristics, to which I will add a fourth one. I will make some specific references to comparisons across sectors and periods in this section, with a particular focus on infrastructure sectors in the nineteenth and twentieth centuries. These types of comparisons are not fundamentally different from others, but they have some particularities that will be highlighted.

#### 2.3.1 Differences in type and style of sources

Any historian using a comparative approach may be confronted with a plethora of sources varying markedly in type and style. Regardless of whether it is a comparison across sectors or periods, the diverse character of sources (and specialist literature) could have an enormous impact on the findings.

Let us consider one possible scenario for a comparison of political systems, international relations or processes of governance across periods: during the nineteenth century political administrations were still small, and documents were only produced by a limited number of administrative staff and politicians. The number of ministries was equally limited. In the 1870s, the Bavarian government for example only consisted of four ministries. The majority of their documents was handwritten, and only one original was produced. In the twentieth century, political systems became much more complex: the number of ministries grew, and the structure of administrations became increasingly layered. These changes resulted in a surge in communications between different administrative entities. In the nineteenth century a single letter to one recipient was often still sufficient to achieve the desired result, but several decades later a similar letter would have to be sent back and forth through the administrative machinery as an increasing number of officials wanted to comment and have their voice heard. In the nineteenth century people only corresponded by handwritten letters, while in the twentieth century non-archivable telephone calls and other forms of electronic media resulted in considerably different types of sources.

In a comparison across sectors, researchers are confronted with similar problems. For instance each sector has its own model and style of preservation of archival documents as well as a particular character regarding policy, politics and polity. In the infrastructure sectors for example, experts often play a decisive role in the decision-making process, but in other sectors such as foreign policy this is often not the case. Differences such as these have an enormous impact on sources, particularly in international comparisons, as national styles of preservation differ markedly. Languages are an additional problem.

It is the researcher's role to assimilate the different source types and styles if at all possible. These challenges entail further problems concerning the selection of sources. All things considered, the choice of sources will have an enormous impact on the findings. Any researcher will therefore have to consider carefully whether similarities and differences in the findings are caused by the type and style of sources and literature. Issues connected to different source types and styles accumulate as comparisons are made across both sectors and periods.

### 2.3.2 Contextualisation and isolation of elements

One of the most crucial aspects of the comparative approach to history is the necessity to isolate a historical case from its context. The elements to be compared usually need to be isolated, causing an interruption in the narrative, in order to define 'comparable units' and consolidate them for analytical purposes. This approach is controversially discussed in the relevant literature because interdependencies in historical developments or between different phenomena cannot be included into the comparison. The intention of this chapter is not to prejudice readers in this matter, but it is totally justified to ask the question: is it really possible to isolate an issue from its context without changing its character? The concept of 'ceteris paribus' – i.e. all variables except the one under study are controlled in order to analyse the effect of a single variable on a dependent variable – is hardly acceptable for many historians (Mahoney/Rueschemeyer 2003, 3).

In historical comparisons, it is advisable to treat historical cases as both 'units of comparison' and 'pieces of a larger jigsaw'. I strongly advocate the isolation of elements (even 'ceteris paribus') within historical comparisons. However, the findings of a comparison will have to be carefully reflected against the context. The benefit of this approach is that the 'bigger picture' is critically examined. Traditional interrelations and connections are questioned, and the path is cleared for new perspectives and forms of explanation beyond the established ones (see paradigmatic purpose of comparisons). Contextualisation is even more difficult if comparisons cover different periods because there is an increased risk of differences in the interrelation with the context. For example the characters of political systems in the nineteenth and twentieth centuries differ completely, but the systems' individual elements or organs can fulfil the same functions in the governance process. Hence it is important to not only take into account absolute similarities and differences but also 'functional equivalents'. Usually contextualisation is more important when trying to establish differences.

The governance of different sectors within the European Economic Community after its creation in 1958 is a very good example for illustrating the influence of the historical context. When comparing governance in the domains of transport and agriculture – two political sectors subject to EEC legislation – it is very easy to come to the conclusion that agricultural policies were very successful while in the transport sector hardly any progress was made. If internal EEC processes are emphasised too strongly in this example, an important contextual variable required to explain policy outcomes is overlooked: the dense network of international organisations that already existed and carried out international governance in the domain of transport. Nothing comparable existed in the agricultural sector. This example demonstrates how a different context can impact enormously on the findings.

### 2.3.3 Focus and definition of boundaries

If comparable units need to be isolated from their context, the parameters of the isolation – the focus and definition of boundaries – come to the fore. It is a fundamental characteristic of comparisons that constructed cases reflect the aims of the research project and purposeful boundaries are defined. The range of time, geography and contents for example can vary considerably. The definition of these boundaries and the selections made will influence the findings. A comparison of international relations in the nineteenth and twentieth centuries from the perspectives of the United Kingdom, France and Germany will produce different results than the Swedish, Estonian or Spanish perspectives. The same applies to time boundaries. When should the period of comparison ideally start and end? How long should it last?

Once again, the more cases included the more important it is to make selective decisions regarding viewpoints, questions and problems. One of the crucial points of the comparative approach to history is to find a suitable level of selection, abstraction and decontextualisation. When making a generalising comparison, any historical case or its elements should also be representative of the entire subject.

### 2.3.4 Parameters of comparison

In addition to the definition of boundaries, comparative parameters and a clear 'tertium comparationes' are characteristic of historical comparisons. It can be a difficult task to establish these. Often theories or models are a good starting point to help researchers define parameters of comparison which will enable them to structure and analyse historical cases. It is nevertheless important to underline that parameters of comparison are merely analytical tools. It is very hard to trace them in the sources. A theory borrowed from the social sciences is usually extracted from contemporary cases. It therefore only partially applies to cases taken from the past, which makes it even more difficult to adopt for comparisons across periods. In historical comparisons, theories and modals require adjustments which may result from the findings of the research process.

## 3. Conceptual Challenges of a Comparison across Sectors and Periods

In the previous chapter we discussed historical comparisons in an abstract way. Now we will shift the focus towards a more pragmatic view on comparative research projects. This chapter reflects experiences with a number of comparative research projects and lessons learned, and should be seen as advice only. It is based on the assumption that a comparative research project consists of three parts which are highly interdependent and not clearly separable: developing a research design, the research process itself, and compiling the results.

### 3.1 Developing a Research Design

The first step of a comparative research project consists of establishing a (preliminary) research design. It can fulfil different functions and be drawn up at very different levels. The initial research design is often forwarded to a supervisor and/or funding agency.

Just like any research project, historical comparisons should start with a definition of the research aim. What is it we want to find out? Is a historical comparison the most suitable approach to reach this aim? In a second and equally important step we should ask: what is our purpose in making a comparison? Which type of historical comparison is suitable for our aim?

Once researchers have answered these questions, they will have to develop a research project by continually reflecting the four characteristics of a historical comparison. This is a challenging part of the research process because previous knowledge of historical cases and developments is required. Researchers have to acquire a 'consciousness of the possible' in order to define cases suitable for comparison (Welskopp 1995, 361). A comparison of many sectors may be interesting, but does it have any scientific value? Is it enough to compare three cases to achieve the desired aim or is it advisable to investigate more (even considering the risk of increased contextualisation problems)?

Before starting the in-depth development of the research design, researchers will have to make themselves familiar with the current state of research and the different lines of discussion within (historical) disciplines. This applies particularly if researchers propose a comparison across sectors, periods and geographical entities where contextual differences play an even more prominent role. This initial work usually generates new questions and ideas. It is obvious that the sources need to be checked in this context (Kaelble 1999, pp. 115–19). From the early stages, researchers should be familiar with differences between source types and styles. The political map of Germany for example differs markedly between the nineteenth and twentieth centuries. Archival sources for the nineteenth century need to be accessed in many different archives, while most resources dating from the twentieth century are kept at the National Archives (Bundesarchiv). In many European countries the ministries of foreign affairs have their own archives. These are separated from the rest of the government archives and are subject to their own styles of preservation and rules of access.

Researchers should also keep in mind that the number of selected cases will determine the extent of archival work. A larger number of cases requires a lot more archival research, but at the same time the number of sources analysed in a project will have to be limited. One of the first steps in developing a research design should be to limit the number of cases and sources to a manageable level. Contextualisation, focus and definition of boundaries will have to be continuously reconsidered.

Probably the most challenging aspect of the comparative approach to history is finding a balance between the 'formative influence' of the research design on the actual research (interpretation of sources) and the impartial interpretation and selection of sources. A comparison makes neither sense if the research design is so ambitious that the sources cannot provide an answer nor if it is so weak that it contains hardly any categories of comparison. Therefore the process of developing the research design requires a variety of reflections and prior knowledge of the sources, the historical case and its context. Preparing a comparative research project – especially across sectors and periods – is time-consuming. Researchers also will have to keep in mind that the first version of the research design often sets the course.

Once a research design has been drafted, building a research hypothesis may be a helpful tool to start the research process. Still, all the above mentioned elements and steps towards a research design depend on pragmatic considerations such as language, staff, finances, access to archives etc.

## 3.2 The Research Process

The development of the research design is followed by a period of ongoing work in accordance with the research design. The transition into this second period is rather fluid. The research process is also a conceptual challenge, although in the literature it is often hidden behind a veil of methodology and terminology. A research design drafted in the development stage is not irrevocable and can still be modified during the research process (apart from the testing of a theory). Instead, the research design and hypothesis should be flexible instruments that can be gradually adjusted when the ongoing research delivers preliminary findings. At this stage, various adjustments may be necessary, and the cases' boundaries may need to be redefined. If a typology is created, subsequent adjustments are at the core of the research because the initial research design and the preliminary typology need to be further refined. Without adjustments of the research design empirical research runs the risk of being used as a convenient support for theoretical models instead of helping to answer the central question. Researchers therefore have to be open-minded throughout the whole research project. Adjustments are of vital importance in comparisons across periods because of contextual differences. Categories of comparison suitable for the twentieth century often turn out to be problematic when they are transferred to the nineteenth century. Still, researchers need to be aware that even small adjustments may have a marked impact on the research results.

Isolating elements and defining boundaries are challenging aspects of the work in the archives. When searching for and interpreting historical sources, the research aim has to be the sole focus. Any aspects connected to the case, but irrelevant for the question should be disregarded. In comparisons across periods, this often feels like walking a tightrope because period-specific characteristics are fundamental for understanding the sources and the context. This is even more challenging for researchers lacking experience in archival work.

The human factor is both a key element and a major conceptual challenge in a comparative research project, especially with a growing complexity of the project. The more cases and researchers included in the project, the more important communication is within the research team. One researcher can very easily adjust a research project following new insights, just by sitting at their desk. In this case the adjustment is limited to a single researcher's decision. In an increasingly complex project and a larger research team, different people with different experiences working on different cases need to cooperatively reflect on any adjustment of the research project. Large groups of researchers on different academic levels (PhD candidates, post-docs, professors etc.) with diverse scientific backgrounds shape the interpretation of sources, and each contributes individually to the adjustment of the research project. They may interpret the same parameters of comparison and the same sources in very different ways. What is an international standard in an infrastructure sector? This question regarding a parameter of comparison may produce hugely different answers, and if the task includes defining a standard for the nineteenth and twentieth centuries, answers would differ even more. In the process of communication a variety of different evaluations of the same parameters will emerge. To make a research team speak the same language and to create a common understanding of the parameters of comparison is a formidable task. Efficiently adjusting a research project or producing valid data and results is very difficult if the research team is too heterogeneous. Interdisciplinarity – which is often considered as the ideal way of research – is therefore an extraordinary challenge for comparative projects.

Communication (both bottom-up and top-down) is vital in a research team using the comparative approach and has a huge impact on the outcome. From the outset, workshops and other forms of communicative interaction should therefore be an integral part of any research project. It may sound obvious, but the cohesion of the research team – ranging from offices to interpersonal relationships – can have an enormous impact on the results of comparative research. In a large team, this approach only makes sense if the members are willing, and able, to contribute to the communication process. Organising the research process in a way that enables an efficient adjustment of the research design is an important – and often underestimated – conceptual challenge of a comparison across sectors and periods.

### 3.3 Compiling the Results

The last phase of a comparative research project starts with a final agreement on the research design. Once a final decision is made on categories of comparison and boundaries, no further adjustments are advisable. As a matter of course an initial research design is never an optimal solution. In most cases it can be improved in one way or another. This applies to comparisons across sectors and periods in particular, where the complex issues inherent in the contextualisation, definition of boundaries or categories of comparison are more important. During the research process, it is nevertheless vital to quickly reach an agreement on adjusting the research design. This is a precondition for a targeted final analysis of the empirical data. In a large research project with individual sub-projects, which include comparisons across sectors and periods, compiling the results is a time-consuming process. It requires a lot of discussion among the research team, particularly if it involves the creation of a typology.

It is important to reconsider the research design and the findings in relation to the four characteristics of historical comparisons right at the end of the project. To what extent will the findings become relative if they are linked back to the original context? Were the parameters and boundaries chosen wisely? Was the comparison a good methodological approach for answering the research question? Are there alternative ways of comparison that might have led to better findings? A critical evaluation of the whole research project along these lines is advisable for several reasons: it helps to assess the findings and to improve further historical comparisons. This applies particularly to methodologically ambitious comparisons across sectors and periods.

## 4. Conclusion

When making historical comparisons, historians cannot be 'lazy'. This applies particularly to comparisons across sectors and periods. On the contrary – they have chosen a very challenging and fascinating approach to history which enables them to draw conclusions that are not made on spurious grounds. A comparative historical approach demands very high levels of performance from researchers. It has various benefits, but is by nature susceptible to criticism and attacks. Boundaries and parameters are always open to attack. Different source types and styles are another problematic aspect, which is reinforced in comparisons across sectors and periods. Nevertheless, if a research design involving a comparative approach is developed and adjusted carefully, it provides us with insights into historical phenomena or developments which are difficult to achieve by traditional ways of 'telling history'.

## References

Bauerkämper, A. 2011. Wege zur europäischen Geschichte. Erträge und Perspektiven der vergleichs- und transfergeschichtlichen Forschung, in *Vergleichen, verflechten, verwirren?* edited by A. Arndt et al. Göttingen: Vandenhoeck und Ruprecht, 33–60.

Berger, S. 2003. Comparative history, in *Writing History. Theory and Practice*, edited by S. Berger et al. London: Hodder Arnold, 161–79.

Diamond, J. and Robinson, J. 2010. *Natural Experiments of History*. Cambridge, MA: Belknap.

Haupt, H.-G. and Kocka, J. 1996. Historischer Vergleich: Methoden, Aufgaben, Probleme. Eine Einleitung, in *Geschichte und Vergleich. Ansätze und Ergebnisse international vergleichender Geschichteschreibung*, edited by H.-G. Haupt et al. Frankfurt: Campus, 9–46.

Kaelble, H. 1999. *Der historische Vergleich. Eine Einführung zum 19. und 20. Jahrhundert*. Frankfurt: Campus.

Kocka, J. 2003. Comparison and beyond. *History and Theory*, 42(1), 39–44.

Mahoney, J. and Rueschemeyer D. 2003. Comparative historical analysis. Achievements and agenda, in *Comparative Historical Analysis in the Social Sciences*, edited by J. Mahoney et al. Cambridge: Cambridge: University Press, 3–40.

McNeill, J. 2011. Making Historical Comparisons. *American Scientist*, 98(5).

Paulmann, J. 1998. Internationaler Vergleich und interkulturelle Transfer. Zwei Forschungsansätze zur europäischen Geschichte des 18. bis 20. Jahrhunderts. *Historische Zeitschrift* 267(3), 649–85.

Smil, V. 2010. *Why America is Not a New Rome*. Cambridge, MA: MIT Press.

Welskopp, T. 1995. Stolpersteine auf dem Königsweg. Methodenkritische Anmerkungen zum internationalen Vergleich in der Gesellschaftsgeschichte. *Archiv für Sozialgeschichte*, 35(3), 339–67.

Chapter 3

# Infrastructure and Fragmentation:
# The Limits of the Integration Paradigm

Jens Ivo Engels

Infrastructure research has made important progress during recent years.[1] One of its main features is the concentration on infrastructure's integration effects on local, national and transnational scales. This contribution aims at questioning this predilection, pointing to the limits of the integration paradigm. Though it certainly does not deny the fact that important integration effects may be detected, it calls attention to the opposite effects of infrastructure planning, building and operating: fragmentation and segregation. In this chapter, I will first of all outline the state of the art of infrastructure integration history and some of the historical myths surrounding infrastructures. Secondly, I will present the following arguments: 1. *The limits of integration* points to the fact that integration projects often fail or do not have the intended effects; 2. *The dialectics of integration and fragmentation* shows the inevitable ambivalence of any form of integration be it political, social or territorial; 3. *Infrastructure as a tool for social exclusion, segregation and persecution* reminds us that technical infrastructure, far from serving benign goals, often is explicitly aimed at excluding or suppressing people. The fourth category, *Infrastructure causing conflicts*, highlights the fact that political unrest and friction may be triggered by infrastructure projects.

## 1. Dimensions of Integration – The State of the Art in Infrastructure Historiography

The integrative power of infrastructure seems obvious. Connection is the very nature of transportation, energy and, above all, communication networks. Ralf Roth highlights their capacity of connecting and surmounting frontiers (Roth. Schlögel 2009: 14). This aspect is an integral part of common definitions of the 'network' itself, presented as 'human made, materially integrated structures that cross national boundaries' (v.d. Vleuten and Kaijser 2006: 6).

This vision is closely connected to our understanding of the expansion of infrastructure. Following the economics of scale, network technologies tend to

---

1   I am grateful to Birte Förster for her important comments on a draft version of this text.

grow in order to reduce average costs and to conquer new markets, thus connecting an ever-growing number of users. Last but not least the everyday experience of the electronic media and global transportation networks leads us to identify economic and cultural globalisation with technical infrastructures. Indeed, scholarly research demonstrates the pivotal role of infrastructure regarding historical globalisation (Schot 2007).

Historical infrastructure research has described multiple integration effects of technical infrastructures that I can only sketch here.

Political integration histories have been extended by infrastructure integration histories. The history of the territorial and administrative integration of nation states has recently been enriched by paying attention to the effects deriving from infrastructure planning and building. The importance of transportation for the cohesion of the Habsburg Empire has been shown as well as a centralisation effect concerning French roads and waterways, even during periods of reduced building activities and infrastructure deterioration (Conchon 2008, Helmedach 2002). The representation of the early modern Saxon postal system has been identified as an important factor of ideological integration of a typical composite monarchy (Behringer 2009).

In recent years, the history of European integration by technical infrastructure has caught the interest of scholars. Above all, the importance of the so-called 'hidden integration' has been highlighted, i.e. European integration beyond political organisation and programs (Misa and Schot 2005). This research branch points to the pivotal role of non-governmental organisations and technical experts and their impact as system builders on integration processes (technical, administrative, economic and political). Indeed, European infrastructure development often played a role for the ideology of political integration, so that, for instance, we can analyse the 'negotiation of the European dream by means of and on European roads' (Schipper 2008: 267). In doing so, infrastructure history has identified different integration strategies with uneven success (Henrich-Franke 2008, 2012). One of the most important debates in this context is devoted to the question if the integration processes where politically intended or simply induced by technical and economic reasons–apparently there are many cases of the latter (Hefeker 2007, Schot 2010, v.d. Vleuten et al. 2007: 324).

Needless to say, the processes aimed at aligning technical networks in order to guarantee their technical interoperability play an important role for transnational integration and are objects of detailed research (Ambrosius and Henrich-Franke 2013, Ambrosius, Henrich-Franke and Neutsch 2010). This holds also true for studies on technical experts and their (transnational) communication networks (Kohlrausch 2008).

Political processes are accompanied by cognitive and cultural integration. This highlights the importance of visual representation of a networked Europe via maps, which led the way to 'perform European-ness' (Badenoch 2010). Moreover, the European transportation network of railways inspired nineteenth-century bourgeoisie of thinking Europe as a connected space, as a 'space full of possibilities' ('Ermöglichungsraum', Dienel 2009).

Recently, technology has been identified as a key factor of worldwide politics (Skolnikoff 1993) or even as a 'prime mover of globalization' (Smil 2010). In particular, recent information technology is hold for pushing political integration or at least convergence. 'Three broad areas of international political life are being altered by digital information technologies: the character and extent of democracy and citizenship, the nature of global political authority, and the nature of security (both for individuals and states)' (Herrera 2003: 533). However, the specific impact of infrastructure compared to technology in general is left to future consideration.

Another branch of infrastructure research focuses on the social integration of modern society by infrastructure. From this perspective, infrastructures are connecting urban and rural spaces, they tend to normalise and to align the needs, the taste and day-to-day practices of the population (van Laak 2001).

However, the dark side of infrastructure has been called into attention repeatedly. The inter-connectedness and technical integration of modern (or modernising) societies implicates specific forms of vulnerability. As infrastructure becomes critical to modern 'networked societies' their breakdown cannot be afforded (Verbong and v. d. Vleuten 2004). Not surprisingly, modern terrorism benefits from the opportunities given by transportation systems as they become crucial, such as the case of the railway since the nineteenth century (Schenk 2010).

All of these studies are principally concerned with the integrating effects of infrastructures. Many of them, it is true, are pointing at the opposite effects, but merely in an anecdotic, not in a systematic way. In the end, actual infrastructure research tends to confirm a vision of technical networks which seems very close to what can be called the historical ideology of infrastructure building.

## 2. Myths and Ideologies of Integration

Besides the actual integration *effects* of infrastructure, it is, above all, the *myth* of integration which seems to be very forceful. Infrastructure has been promoted for a long time based upon the assertion it could foster consensus, inclusion, integration (and also progress, welfare, economic development and prosperity – the latter aspects being not in the centre of interest of this contribution). Most of these myths have been historically developed since the nineteenth century (van Laak 2004). They apply to the international, the national and the local or urban integration.

One of the classical and long-living discourses on infrastructure was inspired by the early nineteenth century utopia sketched by Henri de Saint-Simon and his disciples. He suggested that a new universal, supra-national regime based on infrastructures could be established. Technology and industry would be the basis of a new social and political order. Peace, stability, international cooperation, social integration and the rule of rationality would be the positive effects of this infrastructural regime (Mattelart 2000: 15). Saint-Simonian visions had long-lasting effects on pan-European rhetoric during the nineteenth and twentieth centuries (v. d. Vleuten et al. 2007).

Infrastructure did not only inspire internationalism and universalism. During the period of technocracy the myths of infrastructure were primarily connected with the regulating practices of the nation-state. As Dirk van Laak has demonstrated, technical (and social) infrastructure was regarded as one of the most powerful tools of social engineering. During the so-called 'classical period' of modern infrastructure, i.e. 1880–1970, technological networks seemed to provide technical solutions for social problems, allowing the remodeling and integration of society. Seemingly apolitical, infrastructure promised a better future for all parts of society via technological and economic improvement, implementing social justice and equality, providing supply guarantees on a comparative level for all, ultimately overcoming conflicts and segregation (van Laak 2001: 379, van Laak 2006: 168).

In the specific context of early twentieth century Germany, planning infrastructure was closely linked to visions of the perfectly integrated 'Gemeinschaft' that would overcome serious conflicts, contrasting with the realities of an atomised (modern) society. Organic metaphors used by the planning experts became the cultural symptom of this thinking (van Laak 2001: 386). Moreover, infrastructure development in the colonies meant the connection of remote areas and 'primitive' societies to civilisation and modernity via technical infrastructure – this was at least the idea of its European promoters (van Laak 2004: 408).

With respect to urban development, scholars have debated the outlines of what has been called the 'modern infrastructural ideal' and its decay. This ideal is a set of principles connected to infrastructure from the mid nineteenth to the mid twentieth century. According to this set of principles, urban networks were planned and operated as standardised, monopolised, 'bundled' supplies of energy, water, transportation and communication facilities. Connecting all residents on the same level of quality and costs, it had produced an integrated urban space. Since the 1980s, however, neo-liberal privatisation policies have caused 'splintering urbanism' by unbundling the urban facilities, thus provoking the emergence of gated communities and, 'premium networked spaces', while leaving the rest of the city behind. The modern integrated urban fabric was replaced by a fragmented urban space reinforcing social segregation (Graham and Marvin 2001). However, serious critics have been raised in historiography and (post-)colonial studies, stating that the modern infrastructural ideal had never been fully applied, with the exception of Western metropolises: 'The clash between the modern infrastructural ideal and the contemporary era appears to be largely unfounded. To a large extent, it seems rather to rest on an artificial rapprochement between, on the one hand, a historical model largely confined to the higher-income countries of Western Europe and North America and, on the other, contemporary processes generally observed in cities in lower-income countries' (Coutard 2008: 1819, similar in McFarlane and Rutherford 2008).

To sum it up, the idea of integration by infrastructure has a long and partly problematic history, accompanying infrastructure development since the times of early industrialisation and modern state building. Moreover, it is still debated and, at least indirectly, seriously questioned by the critics of the 'splintering urbanism' paradigm.

## 3. Infrastructure and Disintegration

Of course the legitimisation of infrastructure is based on supposedly positive effects like progress and integration. It is not likely that a promoter of infrastructure presents it as an instrument for social segregation, a cause of conflict and unrest or a means to political fragmentation. However, I intend to show that these effects are an integral part of the history of infrastructures–be it intentional or non-intended.

It is trivial but nevertheless true that the social significance of infrastructures, like any other technical artefact, is not univocal. Rather, it depends on the capacities and needs of each user – and of social practices in general. A stairway may allow the access of many people to a public building, but may exclude wheelchair users (Star 1999: 380).

The connecting qualities of infrastructure, for instance, are not simply technology-induced – they only become reality by human practices. The sheer existence of a technical infrastructure is no guarantee for a corresponding usage. As Geert Verbong has pointed out, trans-border electricity grid connections between West Germany and the Netherlands, once in place, have been used very differently over time, including periods when the two national power systems were literally disconnected because of the small quantity of electricity crossing the border (Verbong 2006).

It is crucial, then, to accept the multiplicity and ambivalence of effects deriving from infrastructure. Of course, infrastructures are 'mediating interfaces' (Badenoch and Fickers 2010), but mediation is limited to those connected: 'Material infrastructure simultaneously connects and disconnects across scales' (McFarlane and Rutherford 2008: 370). Moreover, mediation may imply mechanisms of social segregation or conflict. In the following paragraphs I will outline four categories of effects caused by infrastructures that do not lead to integration, rather to fragmentation, exclusion.

### 3.1 Limits of Integration

My first point does not really concern opposite effects to integration, rather the limits of actual integration processes caused by infrastructure: Integration projects often do fail or do not have the intended effects.

On a macro-historical level we can see that trans-border integration is no linear historical process. The Siegen/Berlin research group on 'Integration von Infrastrukturen in Europa vor dem Ersten und nach dem Zweiten Weltkrieg' has analysed phases of integration and disintegration, caused by economic, political, military and governance influences as well as technological ones (several contributions to Ambrosius, Henrich-Franke and Neutsch 2010 – specifically on the interoperability of national railway systems Schiefelbusch and Dienel 2010). The 'splintering urbanism'-debate (cf. above) refers to comparable phenomena in urban space.

Concerning the above mentioned integration effects of modern electronic media (internet, Twitter etc.) there is of course much more information flow around the world, seeping even into the backyards of dictatorial regimes. Anyhow, as yet no effective transnational political public space has been created, neither on a global level nor even in the European Union; transnational political movements are weak. As Geoffrey Herrera states, the actual effects of technology are very difficult to calculate and the nation state will remain an important player shaping the configuration of information technologies (Herrera 2003: 592).

Although we know a great number of success stories, the history of infrastructure is full of failing integration projects on the micro-level. In 1832, a group of French Saint Simonian engineers tried to implement a national plan for transportation infrastructure. Their goal was a reorganisation of the national territory including the economic integration of peripheries. This scheme failed mainly because of competing regional interests (Hirsch 2003). A hundred years later, French politician Albert Thomas lobbied for a European highway system, which equally had no real chance (v. d. Vleuten et al. 2007).

Even when technically implemented, intended integration effects may fail–as it was obviously the case with the railways in tsarist Russia: it was supposed to serve as a tool for the implementation of 'law and order', political integration and the regulation of social behaviour, fostering 'civilised' manners and orderliness, according to what Petersburg elites regarded as West European standards. In fact, trains and railway stations were differently appropriated by the users and came to be spaces for unrestricted consumption of alcohol and drunkenness. Moreover, the existence of railways helped developing social unrest and political opposition against the tsarist regime, including terrorism (Schenk 2009). A similar case is the history of the GDR highway network. It did not match the expectations of the population because of bad road conditions and failed to rally the inhabitants around the government that intended to create a socialist identity in Eastern Germany (Dossmann 2008: 163–4). Urban water and power networks in twentieth century Los Angeles have been integrated under public ownership for several decades, enabling in principle universal access. However, they did not prevent socio-spatial fragmentation of the metropolis, because the authorities were not able to expand it to the territories of suburban sprawl, thus causing differences between the connected and the unconnected (MacKillop and Boudreau 2008).

Looking the other way round, taking the perspective not from the technical network but from the impacts of 'surrounding' power relations on technology, we can observe similar effects. The Mafia in Sicily is an example for historically extremely persistent and inclusive socio-political power relations. The Mafia's structure and its specific economic strategies, however, inhibited the creation of an integrated water infrastructure, although the national government had intended and subsided it, resulting in a 'fractured and fragmented hydraulic techno-natural edifice' and structural water scarcity (Giglioli and Swyngedouw 2008: 410).

### 3.2 The Dialectics of Integration and Fragmentation

My second point concerns the inevitable ambivalence of any form of political integration, social inclusion, and territorial linking. All of these are dialectically bound to their opposite – which is finally an epistemological statement. To put it simply: inclusion of one group or the connection of one territory means exclusion of the rest. This aspect seems to me the most important argument on infrastructural fragmentation, although it may apply to any political and social process of integration/fragmentation. With respect to technical infrastructures, the dialectics of integration and fragmentation have at least two dimensions: the political/social and the spatial. Of course, these effects are often intertwined.

The political ideology of nationalism is a lucid example for this – it creates exclusions *intra muros* in the given nationalised society, but it cuts also the linkages to neighbouring nations. Technical infrastructures were integral part of this.

One of nationalism's general features consists of a double effect of the inclusion of a more or less heterogeneous ethnic community and the exclusion of other 'alien' groups. This holds true for political processes in a broad sense and especially for the right of citizenship in the nineteenth and early twentieth century (Gosewinkel 2001). Social integration is an ambivalent process which often reproduces social inequalities, as has been shown at the example of German social clubs open to working and middle class members (Nathaus 2010). With respect to infrastructures and the specific case of national socialism, Dirk van Laak shows how the Nazi regime intended to replace the concept of liberal fundamental right by the right of access to (infrastructure-based) services of public interest (van Laak 2006: 175). This policy should enable the 'organic' integration of a seemingly fragmented society – but, needless to say, excluding certain groups, above all Jews and other ethnic groups considered not 'Aryan', and with dramatic effects.

The effects of political nationalism, or rather of the construction of the nation state, on transport infrastructure have been analysed repeatedly. The downfall of the multi-ethnic empires on the Balkan Peninsula in the early twentieth century, followed by the foundation of a number of nation-states, had the effect that the national railway companies tended to isolate themselves. Obviously, the conception of individual national transportation schemes required a policy of intended isolation or, at least, extremely limited cooperation with the neighbouring countries (Tympas and Anastasiadou 2006). Nearly in the same way, the politics of the Cold War put an end to many connections between Western and Eastern Europe, fragmenting the European transport (and power) networks (Trischler 2009).

From the perspective of spatial effects Olivier Coutard has asked the critical question: 'from what spatial scale of universal distribution of a network can one, and should one, consider that its integration effects (within the zone it serves) dominate its fragmentation effects (at its outer edges)?' (Coutard 2008: 1817)

Of course there is no answer to this rhetorical question. Rather, it points to a fundamental conclusion: Processes of integration are co-producing processes of fragmentation and exclusion.

The spatial dimension shows phenomena of territorial fragmentation. Once again, we can observe effects inside and outside a given territory: Infrastructural centralisation leads inevitably to the emergence of peripheries. The fact of being connected or not being connected can divide a society; connectedness and disconnectedness may decide about the political, cultural and economic participation opportunities of the population of a certain area. At the time of the GDR, many East Germans took the opportunity to watch West German TV programs in order to get political information free of socialist propaganda. In a certain part of Saxony people could not receive these programs for topographical reasons. Pityingly they were called inhabitants of the 'Valley of the Ignorants' (*Tal der Ahnungslosen*) (Müller 1991). At present, the so-called bypass strategies of privately owned infrastructure (Graham and Marvin 2001) exclude huge areas from the modernisation of communication infrastructure if it does not seem profitable – even in Germany high speed internet is not available in many rural areas (see Fiedler 2002).

Infrastructure and economic development are closely linked. Modern transportation systems enabled industrialising countries of the nineteenth century not only to establish a division of production between different regions, but fostered also the emergence of economically peripheral regions (Nolte 2009: 135). With respect to the structures of transport in post-1989 Europe, cohesion comes along with centrifugal tendencies (Schlögel 2009: 44); those regions, which get not connected, get 'lost' to modernisation and development. Even the surrounding areas of transit routes are hardly gaining anything, but have to cope with negative by-effects of increased trans-European traffic, such as prostitution (Kaschuba 2009: 183). The same effects have been described with respect to the expansion of the French railways under the auspices of the famous Plan Frevcinet of 1879, when secondary lines were built all over the country. Rural communities connected did profit enormously, whereas settlements without railway stations were condemned to insignificance (Beck 1987).

To sum it up: one of the most important challenges of infrastructure research is to bear in mind the inevitable double effect of integration and fragmentation. Empirically, each case will be characterised by a specific relation of the two effects, giving prevalence to one or the other, but never being without one of the two. Especially the territorial excluding effects of any technical infrastructure are of high importance.

### 3.3 Infrastructure as a Tool of Social Exclusion, Segregation and Persecution

As Langdon Winner has stated in his seminal article, artefacts have politics (Winner 1985). Territorial fragmentation and social exclusion are not only semi-intentional side-effects of infrastructure development which is undertaken to realise inclusion. Rather, they can be an integral part of governing technologies aiming at social exclusion, segregation and simple suppression or persecution.

The clearest examples here is the role of transportation infrastructure during the holocaust, the latter's 'industrial' character deriving not the least from highly complex use of railways and other technical infrastructure (Hilberg 1981). Scholars have retraced the working of integrated coercion-extraction systems including railway, power grids, factories, forced labour and extermination (Maier 2006, Roth 2009). All these practices and technologies were connected to the above mentioned dialectics accompanying the constitution of a German 'Gemeinschaft' inspired by Nazi ideology. To a lesser extent murderous, but nonetheless brutal proved the use of infrastructure as 'tools of empire', enabling the administration, exploitation and military consolidation of European colonies (Headrick 1981, Davis, Wilburn and Robinson 1991).

Of course, infrastructure for the most part has rather 'soft' effects. Langdon Winner (Winner 1985) cites the case of early twentieth century urban planner Robert Moses providing the roads of Long Island with low bridges, preventing public transport to pass and thereby keeping the lower classes off. In general, public transport was a site of racial discrimination and segregation in South Africa and North America. As we have learned from technology studies, this means not only that socially and politically established discrimination was simply mirrored by technical infrastructure, but it was produced by it.

Sanitation policy that for a long time has been regarded as a completely benign project proved to be an instrument of segregation and social as well as racial fragmentation. Practices and discourses of the sanitary city in colonial Bombay drove, above all, a sharp line between western elites and those regarded as uncivilised – i.e. most of the indigenous population. The implementation of sanitation infrastructure often was accompanied by police measures and coercion. 'It tended towards logics of segregation, demolition, regulation, flight, and partial upgrading of infrastructure' (McFarlane 2008: 432). The same can be said about water supply in colonial Batavia: '90% of European residents were connected to the network, and used up 78% of the city's domestic supply, while comprising only 7% of the urban population [...] The new water supply system enabled a specific set of material practices related to indoor hygiene that were coded as signifiers of a racially pure European identity' (Kooy and Bakker 2008: 381–2) whereas the indigenous population had no chance to 'improve' their hygiene because they had no access to these facilities. In turn, everyday practices of Europeans and indigenous people, mediated via water supply infrastructure, seemed to confirm the prejudice of racial as well as cultural differences. It reposed on the binary code of modernity and backwardness and reinforced social-racial segregation.

All in all, infrastructures can be intended to suppress and to exclude people, ranging from brutal persecution to very sophisticated and covert techniques of domination.

## 3.4 Infrastructure Causing Conflicts

Planning, building and operating of large technical systems like infrastructures require coordinated actions by a multitude of organisations and individuals. In fact, these organisations and individuals presuppose and strengthen the integration of economic, technical, administrative, and political agencies. Nevertheless, infrastructure planning and building can trigger the opposite, namely conflict, opposition, and resistance. There is some evidence that it is worth regarding technology as a field structured by tensions and conflict. Conflicts may integrate or disintegrate, strengthen or weaken the bonds between conflicting groups within a given society (Hård 1993: 419). For instance, opposition may turn out to have an integrating effect on a local or sectoral level.[2] However, these effects are limited to comparably small-sized groups; the global effect of conflict consists in the opposite: fragmentation caused by the project in question.

Of course, not every infrastructure scheme causes discontent. However, the larger a project or the higher the number of persons affected, the more resistance is likely. In fact, infrastructure projects in urban contexts 'are produced and contested, and […] can become sites of negotiation, tension and struggle between a variety of interest groups' (McFarlane and Rutherford 2008: 366).

The history of colonialism provides some illuminative examples. In the French Equatorial Africa and French Cameroon of the 1920s, a road-building scheme did not lead to the expected integration. Rather it opened the way for serious discontent: 'Because of the ways they reshaped the region's landscapes and relied on coerced labor from local populations, roads and road work became sites of contestation over colonial rule and order' (Freed 2010).

In their above mentioned study on Batavia/Jakarta, Kooy and Bakker show how the active resistance against the colonist's water supply network by the lower classes has contributed to new, conflicting identities in the indigenous population: Whereas the poor identified themselves with traditional forms of water usage, the middle classes hailed the 'European' water infrastructure as a means of improvement and modernisation. Before the independence from the Netherlands, they blamed the colonists for leaving too many inhabitants unconnected; after the independence they blamed the poor for their unwillingness to accept 'modern' sanitation standards.

One of the most common phenomena in Western societies since the 1970s is the emergence of social movements when planning processes do not take into account the fears and interests of the local population (for environmentalism in Germany cf. Brüggemeier and Engels 2005). Of course, administration and planning experts have learned a lot; more and more they try to anticipate critics and to assure consensus during the planning process (see Haumann 2011). However, there are

---

2   Like many cases of environmental protest, when local or regional populations are developing protest cultures which are closely linked to local identity; several examples in Engels 2006.

still spectacular failures. Recently, a costly project for the rebuilding of the main railway station in the German city of Stuttgart has caused a sort of 'upheaval' in many parts of the town's population. It swept away party loyalties, which had lasted for decades, and paved the way for a spectacular election victory of the Green party in the Land Baden Württemberg. In this case, however, the resistance of the local population was not at all unanimous. Rather, the railway station project is a continuous source of frictions and schisms in the city's constituency.

## 4. Conclusion

Obviously, the political perspective of infrastructure scholars is important for the position they take: On the one hand, (post-)colonial urban studies are regarding infrastructures and related politics of the colonists as being part of imperialistic government practices, privileging negative outcomes and splintering effects. On the other hand, historians of European-ness, many of them being funded by the European Research Council, tend to highlight infrastructure as a non-political merging factor for the continent. As I have tried to demonstrate with the 'dialectics' argument, both are right and both visions are too restricted if they do not take into account the respective opposite effect. Whether integration or splintering *dominates* can be determined for individual cases only and should be elucidated in the awareness of the ambivalent nature of these processes.

Of course my argument is not intended to assert that the integration perspective on infrastructure research is obsolete – nothing short of that. In fact I wanted to demonstrate that the integration perspective is a defying methodological challenge to the scholar taking this point of view. Before assigning integration qualities to a technological system, he or she should at least test the limits of integration, social exclusion effects and possible political conflicts before taking into account the inevitable ambivalence of any possible integration process.

## References

Ambrosius, G., and Henrich-Franke, C. 2013. *Integration von Infrastrukturen in Europa im historischen Vergleich: Synopse.* Baden-Baden: Nomos.

Ambrosius, G., Henrich-Franke, C., and Neutsch, C. 2010. *Internationale Politik und Integration europäischer Infrastrukturen in Geschichte und Gegenwart.* Baden-Baden: Nomos.

Badenoch, A. 2010. Myths of the European Network: Constructions of Cohesion in Infrastructure Maps, in *Materializing Europe*, edited by A. Badenoch and A. Fickers. Basingstoke: Palgrave Macmillan, 47–77.

Badenoch, A. and Fickers, A. 2010. Europe Materializing? in *Materializing Europe*, edited by A. Badenoch and A. Fickers. Basingstoke: Palgrave Macmillan, 1–23.

Beck, R. 1987. Les effets d'une ligne du Plan Freycinet sur une société rurale. *Francia* 15, 561–77.

Behringer, W. 2009. Die Visualisierung von Straßenverkehrsnetzen in der frühen Neuzeit, in *Die Welt der europäischen Straßen. Von der Antike bis in die Frühe Neuzeit*, edited by T. Szabó. Köln: Böhlau, 255–78.

Brüggemeier, F.-J.. and Engels, J.I. 2005. *Natur- und Umweltschutz nach 1945*. Konzepte, Konflikte, Kompetenzen. Frankfurt a.M.: Campus.

Conchon, A. 2008. Les transports intérieurs sous la Révolution: Une politique de l'espace. *Annales Historiques de la Révolution Française,* 352, 5–28.

Coutard, O. 2008. Placing Splintering Urbanism. Introduction. *Geoforum,* 39, 1815–20.

Davis, C.B.. Wilburn, K.E.. and Robinson, R.E. 1991. *Railway Imperialism*. New York: Greenwood Press.

Dienel, H.-L. 2009. Die Eisenbahn und der europäische Möglichkeitsraum, 1870–1914, in *Neue Wege in ein neues Europa*, edited by R. Roth and K. Schlögel. Frankfurt a.M.: Campus, 105–23.

Dossmann, A. 2008. Socialist Highways? Appropriating the Autobahn in the German Democratic Republic, in *The World beyond the Windshield. Roads and Landscapes in the United States and Europe*, edited by C. Mauch and T. Zeller. Athens. Stuttgart: Ohio University Press. Franz Steiner Verlag, 143–67.

Engels, J.I. 2006. *Naturpolitik in der Bundesrepublik. Ideenwelt und politische Verhaltensstile in Naturschutz und Umweltbewegung 1950–1980.* Paderborn: Schöningh.

Fiedler, C. 2002. *Telematik im ländlichen Raum Bayerns. Möglichkeiten und Grenzen zur Minderung von Standortnachteilen in ländlichen Gebieten.* Bamberg: Institut für Geographie.

Freed, L. 2010. Networks of (colonial) power: Roads in French Central Africa after World War I. *History & Technology,* 26, 203–23.

Giglioli, I. and Swyngedouw, E. 2008. Let's Drink to the Great Thirst! Water and the Politics of Fractured Techno-natures in Sicily. *International Journal of Urban and Regional Research,* 32, 392–414.

Gosewinkel, D. 2001. *Einbürgern und Ausschließen. Die Nationalisierung der Staatsangehörigkeit vom Deutschen Bund bis zur Bundesrepublik Deutschland.* Göttingen: Vandenhoeck & Ruprecht.

Graham, S. and Marvin, S. 2001. *Splintering Urbanism. Networked Infrastructures, Technological Mobilities and the Urban Condition.* London. New York: Routledge.

Hård, M. 1993. Beyond Harmony and Consensus. A Social Conflict Approach to Technology. *Science, Technology & Human Values,* 18, 408–32.

Haumann, S. 2011. '*Schade, daß Beton nicht brennt ... 'Planung, Partizipation und Protest in Philadelphia und Köln 1940–1990.* Stuttgart: Franz Steiner Verlag.

Headrick, D.R. 1981. *The Tools of Empire. Technology and European Imperialism in the Nineteenth Century.* Oxford: Oxford University Press.

Hefeker, C. 2007. Die europäische Währungsintegration nach dem Zweiten Weltkrieg: Politik, Ideologie oder Interessen?, in *Internationalismus und Europäische Integration im Vergleich. Fallstudien zu Währungen, Landwirtschaft, Verkehrs- und Nachrichtenwesen*, edited by C. Henrich-Franke, C. Neutsch and G. Thiemeyer. Baden-Baden: Nomos, 57–81.

Helmedach, A. 2002. *Das Verkehrssystem als Modernisierungsfaktor. Straßen, Post, Fuhrwesen und Reisen nach Triest und Fiume vom Beginn des 18. Jahrhunderts bis zum Eisenbahnzeitalter*. München: Oldenbourg.

Henrich-Franke, C. 2008. Mobility and European Integration. Politicians, Professionals and the Foundation of the ECMT. *Journal of Transport History*, 29, 64–82.

Henrich-Franke, C. 2012. *Gescheiterte Integration im Vergleich. Der Verkehr - ein Problemsektor gemeinsamer Rechtsetzung im Deutschen Reich (1871–1879) und der Europäischen Wirtschaftsgemeinschaft (1958–1972)*. Stuttgart: Steiner.

Herrera, G.L. 2003: Technology and International Systems. *Millennium – Journal of International Studies*, 32, 559–93.

Hilberg, R. 1981. *Sonderzüge nach Auschwitz*. Mainz: Dumjahn.

Hirsch, J.P. 2003. Saint-simonisme et organisation du territoire. Sur un programme de 1832. *Revue du Nord*, 85, 861–83.

Kaschuba, W. 2009. Europäischer Verkehrsraum nach 1989 – die Epoche der zweiten Globalisierung, in *Neue Wege in ein neues Europa*, edited by R. Roth and K. Schlögel. Frankfurt a.M.: Campus, 175–94.

Kohlrausch, M. 2008. Technologische Innovation und transnationale Netzwerke: Europa zwischen den Weltkriegen. *Journal of Modern European History*, 6, 181–95.

Kooy, M. and Bakker, K. 2008. Technologies of Government: Constituting Subjectivities, Spaces, and Infrastructures in Colonial and Contemporary Jakarta. *International Journal of Urban and Regional Research*, 32, 375–91.

Laak, D. v. 2001. Infra-Strukturgeschichte. *Geschichte und Gesellschaft*, 27, 367–93.

Laak, D. v. 2004. *Imperiale Infrastruktur. Deutsche Planungen für eine Erschließung Afrikas 1880 – 1960*. Paderborn: Schöningh.

Laak, D. v. 2006. Garanten der Beständigkeit, in *Strukturmerkmale der deutschen Geschichte des 20. Jahrhunderts*, edited by A. Doering-Manteuffel. München: Oldenbourg, 167–80.

MacKillop, F. and Boudreau, J.-A. 2008. Water and Power Networks and Urban Fragmentation in Los Angeles: Rethinking Assumed Mechanisms. *Geoforum*, 39, 1833–42.

Maier, H. 2006. Systems Connected, in *Networking Europe*, edited by E. v. d. Vleuten and A. Kaijser. Sagamore Beach: Science History Publications, 129–58.

Mattelart, A. 2000. *Networking the World 1784–2000*. Minneapolis: University of Minnesota Press.

McFarlane, C. 2008. Governing the Contaminated City: Infrastructure and Sanitation in Colonial and Post-Colonial Bombay. *International Journal of Urban and Regional Research,* 32, 415–35.

McFarlane, C. and Rutherford, J. 2008. Political Infrastructures: Governing and Experiencing the Fabric of the City. *International Journal of Urban and Regional Research,* 32, 363–74.

Misa, T.J. and Schot, J. 2005. Inventing Europe. Technology and the Hidden Integration of Europe. *History & Technology,* 21, 1–19.

Müller, H.G. 1991. *Im Tal der Ahnungslosen. Rückblick auf eine Reise nach Dresden vor der Wende.* Frankfurt a.M.: Haag und Herchen.

Nathaus, K. 2010. Vereinsgesselligkeit und soziale Integration von Arbeitern in Deutschland, 1860–1914. *Geschichte und Gesellschaft,* 36, 37–65.

Nolte, H.-H. 2009. Eisenbahnen und Dampferlinien, in *Neue Wege in ein neues Europa,* edited by R. Roth and K. Schlögel. Frankfurt a.M.: Campus, 124–40.

Roth, R. 2009. Wenn sich Kommunikations- und Transportsysteme in Destruktionsmittel verwandeln – die Reichsbahn und das System der Zwangsarbeit in Europa, in *Neue Wege in ein neues Europa,* edited by R. Roth and K. Schlögel. Frankfurt a.M.: Campus, 235–60.

Roth, R. and Schlögel, K. 2009. Einleitung. Geschichte und Verkehr im 20. Jahrhundert, in *Neue Wege in ein neues Europa,* edited by R. Roth and K. Schlögel. Frankfurt a.M.: Campus, 11–26.

Schenk, F.B. 2009. Im Kampf um Recht und Ordnung: Zivilisatorische Mission und Chaos auf den Eisenbahnen im Zarenreich, in *Neue Wege in ein neues Europa,* edited by R. Roth and K. Schlögel. Frankfurt a.M.: Campus, 197–221.

Schenk, F.B. 2010. Attacking the Empire's Achilles Heels: Railroads and Terrorism in Tsarist Russia. *Jahrbücher für Geschichte Osteuropas,* 58, 232–53.

Schiefelbusch, M. and Dienel, H.-L. 2010. Zielkonflikte und Interessensgegensätze bei der Entwicklung des europäischen Eisenbahnsystems, in *Internationale Politik und Integration europäischer Infrastrukturen in Geschichte und Gegenwart,* edited by G. Ambrosius, C. Henrich-Franke and C. Neutsch. Baden Baden: Nomos, 61–86.

Schipper, F. 2008. *Driving Europe. Building Europe on Roads in the Twentieth Century.* Amsterdam: Aksant.

Schlögel, K. 2009. Europa in Bewegung – Die Transformation Europas und die Transformation des europäischen Verkehrsraumes, in *Neue Wege in ein neues Europa,* edited by R. Roth and K. Schlögel. Frankfurt a.M.: Campus, 29–46.

Schot, J. 2007. Globalisering en Infrastructuur. *Tijdschrift voor Sociale en Economische Geschiedenis,* 4, 107–28.

Schot, J. 2010. Transnational Infrastructures and the Origins of European Integration, in *Materializing Europe.* edited by A. Badenoch and A. Fickers. Basingstoke: Palgrave Macmillan, 82–109.

Skolnikoff, E.B. 1993. *The Elusive Transformation. Science, Technology, and the Evolution of International Politics.* Princeton, NJ: Princeton University Press.

Smil, V. 2010. Two Prime Movers of Globalization. The History and Impact of Diesel Engines and Gas Turbines, Cambridge, MA: MIT Press.

Star, S.L. 1999. The Ethnography of Infrastructure. *American Behavioral Scientist,* 43, 377–91.

Trischler, H. 2009. Geteilte Welt? Verkehr in Europa im Zeichen des Kalten Krieges, 1945–1990, in *Neue Wege in ein neues Europa*, edited by R. Roth and K. Schlögel. Frankfurt a.M.: Campus, 156–74.

Tympas, A. and Anastasiadou, I. 2006. The Modern Greek Pursuit of an 'Iron Egnatia', in *Networking Europe*, edited by E. v. d. Vleuten and A. Kaijser. Sagamore Beach: Science History Publications, 25–49.

Verbong, G. 2006. Dutch Power Relations, in *Networking Europe*, edited by E. v. d. Vleuten and A. Kaijser. Sagamore Beach: Science History Publications, 217–44.

Verbong, G. and Vleuten, E. v. d. 2004. Under Construction: Material Integration of the Netherlands 1800–2000. *History & Technology,* 20, 205–26.

Vleuten, E. v. d. and Kaijser, A. 2006. Prologue and Introduction, in *Networking Europe*, edited by E. v. d. Vleuten and A. Kaijser. Sagamore Beach: Science History Publications, 1–22.

Vleuten, E. v. d. et al. 2007. Europe's System Builders. The Contested Shaping of Transnational Road, Electricity and Rail Networks. *Contemporary European History,* 16, 321–47.

Winner, L. 1985. Do Artifacts have Politics? in *The Social Shaping of Technology*, edited by D. MacKenzie and J. Wajcman. Milton Keynes. Phiadelphia: Open University Press, 26–38.

# Chapter 4

# Archetypes of International Infrastructural Integration

Gerold Ambrosius

## 1. Introduction

This chapter endeavours to answer the question whether archetypes of infrastructural integration between countries exist. It is needless to say that cross-border traffic of infrastructural goods and services and cross-border links between infrastructural networks are vital for economic development. Infrastructures are the basis for most types of economic activity, regardless of whether they span cities or regions, national economies, the European economic area, or even the world economy. Economic integration in general and infrastructural integration in particular are interdependent. The more goods and services, labour and capital cross national borders, the higher the demands on infrastructures and how they are linked across borders. The stronger the links between national infrastructures, the more products and factors of production will be exchanged between countries. This interdependence is not just a recent development, but historically has always existed. In addition to a systematic or theoretical perspective, this chapter will therefore take a historical-empirical approach to answering the question of archetypes. If there is an empirical element to the study, there are only archetypes in terms of time, geography and sector. With this in mind, in this chapter the term archetype will be used as an 'ideal type' based on historical precedent rather than an abstract and theoretical construct.

In the following sections, we will first explore the meaning of the term infrastructural integration. Then we will outline the main reasons for infrastructural integration, and finally attempt to identify archetypes. At the beginning of each section, we will consider economic integration and the integration of economic policies to illustrate how closely both correlate to infrastructural integration.

## 2. What Is Infrastructural Integration?

In this chapter, the term economic integration will refer to both the increasing interrelations between the markets for goods and services, labour and capital, and to the convergence of economically relevant parameters, standards and norms. Parameters include for example interest rates, exchange rates, prices, wages, or per capita income. Standards are technical, operational, legal, tariff and other widely applied requirements. Norms are established social patterns of behaviour such as conventions, manners or values. It is very difficult to clearly define or distinguish parameters, standards or norms. The relationships between their convergence and the interrelation of markets are equally elusive: convergence may or may not be conducive to interrelations. In this chapter we will mainly look at standards (De Vries 2002).

The integration of economic policies defines the convergence of territorial or national economic policies, in this case meaning government-led policies. Convergence of policies primarily refers to politics, but it may also apply to policy and polity. Convergence in one of these areas is not necessarily accompanied by convergence in the others. The interrelation of markets and the convergence of standards in the market (economic integration) and the alignment of policies including standards (integration of economic policies) are often interdependent, but they do not need to be. It is very difficult to precisely identify the policies relevant for economic integration, because nearly all of them are relevant in some way, including the alignment of standards (Molle 1991).

Standards commonly used within national economies are harmonised across borders by means of standardisation. In this respect national standards can be standardised internationally, and the terms standardisation and integration are used synonymously in this context. Standardisation or integration can be agreed 'privately' by economic entities or enforced by government policies. It is not a state, but a process.

Economic infrastructures generally include various sub-sectors of transport (rail, shipping, road, aviation), communications (postal system, telegraphy/ telephony, radio) and public utilities (water, gas, electricity, oil). The term applies to both networks and services. Economic integration in general and infrastructural integration in particular can be described as a growth in cross-border transport of material goods and immaterial services.

However, as this type of transport frequently requires dedicated networks, the increasing links between these networks intended by standardisation are also part of infrastructural integration. In essence it is a question of whether these networks or 'systems' are compatible, connectable and combinable. The aim is to achieve better cooperation at interfaces (interconnectivity), the consolidation of whole systems (interoperability) and the possibility to transfer single components (portability) (Blind 2004).

## 3. What are the Reasons for Infrastructural Integration?

If we define economic integration as an increasing interrelation between markets and alignment of standards, distinguishing between causes and objectives only makes limited sense. From an economic point of view, the most frequently quoted objectives of economic integration are a more efficient allocation of resources, a more varied and lower-cost offer, higher economies of scales, more vigorous competition, and consequently increased rationalisation and accelerated product and process innovation, and a reduction in various transaction costs – in short: stronger growth and greater prosperity (Blind 2002, Swann 2000). This last point illustrates clearly that it is very difficult to separate objective and cause: increasing prosperity is an objective of economic integration, but history has shown that it can also be its cause.

For the integration of economic policies, causes and objectives are partially identical to the ones just mentioned. The integration of economic policies is intended to promote economic integration in order to increase prosperity. National governments could exploit the integration of economic policies to distract from their own responsibility and shift the blame for awkward decisions to an international level. On a positive note, the integration of economic policies may support overriding objectives such as peacekeeping, promoting democratic developments or reinforcing social or cultural bonds across borders. While economic integration will indirectly help bring about these political and social objectives, economic policies often contain an explicit statement of intent to this effect. At least this is the publically delivered message.

Infrastructural integration is ultimately based on the same underlying causes and objectives, with possibly additional ones arising from the particular economy of infrastructural networks. Accordingly, standardisation or integration results from so-called network effects which may either be direct or indirect (Blankart and Knieps 1992, Weitzel 2000). A direct network effect refers to the fact that a single user will not be able to use network goods or services independently of other users. This frequently results in positive network effects, i.e. the benefit of those goods or services may be positively linked to the number of those using them. For example using a specific mobile phone network makes more sense if there is already a large user base. There are also negative network effects which may arise because a large number of network users (installed) vie for the same resources. For example high levels of road traffic will lead to congestion. But negative effects can also facilitate integration if they result in a network expansion (Katz and Shapiro 1985, 1986 and 1992, Hofmann 2001).

Indirect network effects describe compatibility effects, learning effects and expectancy effects (Glanz 1993, Knorr 1993, Nicklas 2000, Thum 1995):

(1) Compatibility effect means that with increasing technical similarities the number of possible combinations between products and processes rises, which both users and providers will appreciate. Users will profit from a broader network base which allows for a growing number of complementary products and processes, and this in turn will be beneficial for all users. For providers, increasing compatibility opens up the prospect that users will come to appreciate their products over those of other providers. Sales prospects will improve. However, these positive effects may also entail negative ones: if the goods and services offered are technically too similar, providers run the risk of losing custom to competitors.

(2) Learning effect means that the benefits of many new products, processes or technologies are initially unknown, and the question remains to be answered whether they will replace established technologies in the future. Users will have to learn by experience what each new technology is suitable for. The more experience people gain, the more information will be shared which creates an interdependency between the users. Due to these learning effects the value of products or processes increases with a growing user base.

(3) Expectancy effect means that for products or processes with a long life span there will most likely be a demand for spare parts and complementary goods in the future. The more users a product or a process has, the more likely it will be available in the long-term. Each buyer helps increase the likelihood of a longer life span on the market and the potential for further standardisation.

So far we have focused on the technical and economic logic of networks and supplier or customer behaviour. In many cases, this would sufficiently explain what causes standardisation. In others it will not. Historians in particular will need to explore the role of political, social, legal or cultural factors in the standardisation or integration process. Even though the rational calculus is particularly important for economic and technical standards, they can only be explained within the context of specific social circumstances. Incidentally, standards will influence the social environment, and changes in social environment will in turn impact on standardisation and integration (Ambrosius 2009).

## 4. What are the Archetypes of Infrastructural Integration?

In principle, economic integration can take place competitively or cooperatively. Cross-border traffic of goods and services, labour and capital is nearly always based on competition, because agreements between private companies and stakeholders, including international cartels, are primarily intended to apportion markets and prevent integration. When standards are aligned, the relationship between competition and cooperation is different insofar as competition also plays a considerable role in international standardisations, but negotiations between companies and stakeholders are equally important. This will be explained in further detail in connection with the archetypes of infrastructural standardisation described below.

For the integration of economic policies, these two fundamental options of convergence – competition and cooperation – are equally relevant. Competitive integration can be described as follows (Kasper and Streit 1998): in most cases societies or their jurisdictions respectively compete with each other. Employees are after high wages and favourable working conditions, while employers want cheap and skilled labour. Consumers demand affordable, high-quality goods and services. Manufacturers require a favourable production climate, while investors seek out good conditions for profitable investments. All of them operate in a domestic environment, but also internationally. Competition between jurisdictions arises because labour, consumer market, production and investment conditions are largely determined by the relevant institutional framework, for example by the law. This type of institutional competition may or may not lead to a convergence of jurisdictions, policies or standards. If an alignment of standards is achieved, in this case it is not due to cooperation, but rather to competition.

By contrast, cooperative integration can take place as a spontaneous collaboration through political negotiations (Abott and Snidal 2001). In this case the collaboration ends as soon as the parties achieve a specific integration objective. It can also be a result of permanent collaboration in established international organisations. In this context, negative integration means that barriers to trade and mobility such as customs duties or quantitative trade restrictions are removed in a concerted effort (Molle 1991). Positive policy integration means that these barriers to trade and mobility will be lowered by aligning jurisdictions, policies or standards such as directives on labour, consumer or health protection in order to create equal conditions for the operation of the integrated parts of the economies. Integration theory distinguishes two ideal-typical extremes, complete self-sufficiency and complete abandonment of sovereign rights, with various levels of integration in between:

1. preference zone: dismantling of tariffs for specific or all product groups between different countries;
2. free trade zone: exemption from duties within the zone, member-specific duties to countries outside the zone;
3. customs union: exemption from duties within the zone, common duties to countries outside the zone;
4. common market: customs union, free trade of goods and services and unlimited mobility of labour and capital;
5. single market: common market with some degree of consistency in production conditions;
6. economic and monetary union: single market with harmonised economic policies, some supranational bodies and irrevocably fixed exchange rates or a single currency.

Integration of economic policies across borders will ultimately lead to a political merger of the countries party to the integration into a single (federal) state. These general statements on integration policies apply basically also on the integration of infrastructures. But only recently the harmonisation of infrastructural policy has started in the community. Still nor the European Union is lacking effective instruments for an independent supranational policy.

These levels of integration are based on a static view, the ways integration is achieved on a more dynamic one. Concerning this dynamic view integration theory provides us with some ideal types – federalism, inter-governmentalism, supranationalism and others – defined by the extent to which the countries give up their sovereignty to supranational organisations and by the way politics is practised in these integrated entities. This chapter will primarily focus on the standardisation of products and processes of particular importance for infrastructural integration.

## 4.1 How Standards Are Enforced

### 4.1.1 Competition in 'the market'

Standardisation can take place through competition in the market (Besen and Farell 1994, Farell and Saloner 1988, Gabel 2005, Glanz 1993, De Vries 2006a). In this case there are no negotiations and agreements. The major players are companies, and in infrastructural markets often public companies or administrations. Coordination takes place through pricing, convention or imitation, depending on the point of view. Market participants carry out cost-benefit-analyses, and may adopt other companies' products or processes without being innovative. In this way, these products and processes will gradually gain more ground and ultimately be recognised as international standards. The more widely used products and processes are, the more worthwhile it is to adopt them. However, this competitive, non-cooperative type of standardisation will only work smoothly under very restrictive conditions. The potential standard will have to be freely accessible. Market participants will have to be fully informed about their options and share similar interests. There should be no major disparities in distribution, meaning standardisation should not cause serious losses for some market participants, while others achieve substantial profits. If these conditions are met, a chain of imitations will be triggered which will lead to comprehensive standardisation.

The reality on the international infrastructure markets is somewhat different. Competitors may prevent a provider of products and processes essential for infrastructures from reaching the critical mass of users required to attract more. Given the monopoly some of the public providers of services enjoy, smaller countries will struggle to impose their standards on providers in larger countries. The market form is generally of no little consequence to standardisation. There may be delays because entities that would like to adopt a product or process cannot be certain others will follow suit; they run the risk of making the wrong investment. A widespread wait-and-see-attitude may even prevent standardisation from taking place at all. On an international level, national standards for terminal devices for example can be used purposefully as a non-tariff barrier, which will also hamper network integration (Fischer and Serra 2000). The viability of these competitive and uncoordinated standardisations is also limited by the fact that the countries set the legal framework for competitive standardisation on a national level, which may either complicate or facilitate international standardisation.

### 4.1.2 Cooperation in 'the committee'

In committees, standardisation invariably occurs cooperatively through negotiations (Chiao and Tirole 2007, Geschle 1995, Jakobs 2000, Voelzkow 1996, Werle 1997 and 2001). Committees can be loosely organised standard bodies, more formal consortia and forums, or permanently established organisations. Members may include manufacturers and providers, consumers, experts, and in some cases government officials. Information is either disseminated by informal agreements or formal contracts. The purpose of negotiations in a committee is to reach an agreement on standards which will benefit all participants and leave none of them disadvantaged. Following a successful agreement, a standard will only be created automatically if a formal contract is made. However, as there is an underlying consensus that the negotiated solutions will benefit all parties, everyone will expect the agreement to be adopted. It is therefore very likely that the solutions will be accepted as standards even with just an informal agreement (Egyedi 2003).

In negotiating standards, three ideal-typical scenarios with different coordination issues may emerge (Mattli 2001, Mattli and Büthe 2003, Schmidt and Werle 1998):

(1) all parties want to establish the same standard. In this case, the negotiation and coordination of standards will be straightforward, and the parties will swiftly come to an agreement without any problems. If the group of interested parties is large and includes the majority of providers in a particular market, it is very likely that their agreement will prevail throughout the market and not only among the committee members. The same applies to a small committee made up of major providers of infrastructural goods and services. By contrast, in a small group of providers without significant market power such an outcome is unlikely. There is the potential danger that the standard preferred by all participants may not necessarily be the most suitable one. As alternatives are not seriously taken into consideration, a lock-in may occur that is difficult to resolve, precisely because the common standard had been agreed on cooperatively by a committee (Arthur 1989, Farell and Saloner 1985, Spryt 2001).

(2) Standardisation in a committee will evolve differently if all parties want to establish a common standard, but disagree on which one. The extent to which the parties' interests diverge depends chiefly on how recently the coordination issue has arisen: the more recent, the better the prospects for a successful cooperative standardisation, because the major players have not yet acquired much experience with a product or process and they will have invested less in their own technology. Consensual negotiation is both an advantage and a disadvantage of standardisation in a committee. If the preferences for the common standard differ too much, committee members may prevent a consensus, for example because the benefits are not distributed evenly. In this case, the committee will fail to agree on a common standard, at least cooperatively.

(3) Finally, coordination issues may also arise because some parties want a common standard and others do not due to major disparities in distribution. Some may profit considerably from a common standard because it will facilitate the combination of existing technology with complementary technologies (complementary compatibility). By contrast, others may be left significantly

disadvantaged, because their technology may be replaced by another (substitutive compatibility). In any case, an agreement on a common standard will not be reached if committee members oppose it on principle.

Cooperative standardisation is significantly shaped by the major players, structures and processes of the organisations, committees or other entities, in which the negotiations take place. Strategies within the different configurations have been modelled in all possible nuances, using game theory. In addition to formal membership rules, responsibilities and procedures there are informal ones, which are also part of the 'organisational culture'. Beyond the rationality of game or negotiation theory an entire panorama of relevant behavioural options opens up: self-interest and power vs. consensus and balance, ideology and norms vs. intersubjectivity and objectivity, tactics and strategy vs. openness and fairness, lobbyists and politicians vs. scientists and experts etc.

Economic and/or political power evidently plays a particular role in all forms of coordination. Sufficiently powerful providers will not need to bother with cumbersome negotiations in a committee, because they can feel certain that their products and processes will take hold in the market. Even in formal organisations, where standards need to be negotiated, the large providers will have more opportunities to have their own products and processes accepted as standards (Ambrosius, Henrich-Franke and Neutsch 2010, van de Vleuten and Kaijser 2006).

### 4.1.3 Hierarchical order by public authorities

The third ideal-type of enforcing standards is the hierarchical order by public authorities. After a democratic decision-making process the government, i.e. the executive, can enforce and implement standards in a hierarchic-bureaucratic way. They are imposed by legislation, through legal acts and regulations. However, this type is not relevant in the context of this chapter as there was no 'state' at the international level in the past. All the international organisations which occupied with infrastructural integration resp. standardisation were inter-governmental and not supranational. Of course the Union kept on trying to enforce important standards on the member states also on the infrastructural field. But these discussions had mostly the character of inter-governmental negotiations with a strong veto position of the member states. In so far even in the European Union with its supranational approach standardisation is achieved through 'negotiation within a committee'.

### 4.2. Types of Standards

In addition to the ways standards are enforced, infrastructural integration is also influenced by particular types of standards. Standards can be classified according to a variety of criteria (cf. Figure 4.1). There is neither a clear system or hierarchy nor a clear distinction between standards. It all depends on the topic under study. A few types are mentioned below, headed by different criteria

**Types of standards**

| technical co-ordination | portable | interoperational | interconnectiv | compatible |
|---|---|---|---|---|
| accessibility | restricted/proprietary | | open/communitary | |
| external effect | coordinative | | regulative | |
| content | tariff-related | technical | operational | etc. |
| range | company-wide | local | regional | national | global |
| character | public | collective | private | |
| impact | allocative | | distributive | |

**Types of implementation**

| decision-making | actors | medium |
|---|---|---|
| competition in "the market" | enterprises, states | price, imitation |
| cooperation in "the committee" | enterprises, states | agreement, contract |

**Figure 4.1   Types of Standardisation (on international level concerning infrastructures)**

of classification: implementation – de facto or de jure standards; accessibility – restricted or open standards; contents – tariff-related, technical, operational etc. standards; range – company-wide, local, regional, national or global standards; impact – allocative or distributive standards; external effects – coordinative or regulative standards; ownership – public, collective or private standards; type of compatibility – transferable, interoperative or interconnective standards; legal character – legislative or private standards. This list is not exhaustive (De Vries 1998 and 2006b).

To give at least some definitions that could be relevant for historical research: An open standard is publicly available and has various rights to use associated with it, and may also have various properties of how it was designed. A public standard is available for use by all and that use by any one economic actor does not reduce the amount available to others. A restricted or private standard is in possession of somebody and can only be used or diffuse by getting permission or paying a charge. A coordinative standard ensure compatibility between different products or technologies and produce positive externalities. It is a voluntary standard, set by market players and diffuse through market dynamics – also on the international political market with states as market players. A regulative standard describes requirements on a product, process or service in order to establish their fitness for purpose and are referred to as quality standards. It is set by, or under control of, public authorities, and enforced by imposition. In reality standards are of course never wholly private or wholly public, wholly coordinative or wholly regulative, and neither are the ways of diffusion.

Many combinations of these types of standards are possible, but some do not work. For example, a company-wide standard can be technical, coordinative, restricted and transferable, but not collective. An open standard is designed to be used by anyone and may be coordinative and technical, but not private.

From a specific perspective, it evidently makes sense to reduce an analysis to just a few types. Quite a few studies which have delivered valuable insights focused either exclusively on open or restricted, or on coordinative or regulative standards. Some interesting aspects will obviously be neglected in doing so, because most standards are very multi-faceted. All these different aspects apply to both national and international standards.

### 4.3 Types of Standardisation or Integration

In order to identify 'archetypes' of integration in the sense of historical-empirical ideal types, the relationship between forms of enforcement, types of standards and infrastructural sectors has to be established, but the historical context equally needs to be taken into account. Integration of standards takes place in a specific technical, economic, social and cultural environment that will shape its type. Below only some hypotheses will be given in an attempt to identify archetypes. They are based on research of historical experiences with international integration of infrastructures in Europe (Ambrosius et al. 2009, Ambrosius et al. 2010, Auger et al. 2009, Badenoch and Fickers 2010, van de Vleuten and Kaijser 2006).

*Coordinative technical standards spread primarily through competition in the market throughout the nineteenth century, while after World War II they were negotiated in committees.*

This type of standard is generally characterised by positive external effects. Both the operators and the users of a 'network' will profit from standardisation. Their impacts are primarily allocative, while the distributive ones are uncertain. This applies to the nineteenth century as much as to the twentieth century. However, in the nineteenth century some infrastructures were still mixed competitive-monopolistic markets with private and public providers which were at least partially competing against each other. Besides, infrastructural technologies had only recently been developed. As yet, providers had not 'sunk' vast amounts of money in investments. The committees, which already existed at the time, abstained from certain standardisations to avoid curtailing the flexibility of technological progress. Even the regulative framework imposed by national governments hardly interfered with competitive standardisation. By contrast, after World War II there was only a very limited number of public providers, and all of them enjoyed a monopoly. They had long since committed to specific technologies, and made substantial investments. Furthermore, they operated in a regulative environment which subjected even the smallest technological innovation to a strict state control. Besides, since the nineteenth century international standardisation bodies with stable organisational cultures and established decision-making processes had developed.

*Regulative quality and safety standards were negotiated and enacted by committees in the late nineteenth century, and on an international level both government and private entities participated in this process. In the second half of the twentieth century, within the European Community, participation in the standardisation committees became increasingly limited to government entities. It was only recently that private entities became once again more involved.*

In the nineteenth century, the definition of quality standards for food products or safety standards for the workplace on a national level was initially left to private companies, but later subjected to state control. Although private experts and stakeholders were still involved in negotiating the standards, the final decision and the enforcement were increasingly taken out of their hands by government bodies. On an international level, infrastructural safety standards underwent a similar development. Since the mid twentieth century they were negotiated by international committees. Private entities were still involved in the standardisation process as they played an important role in the infrastructural markets. From then on, the importance of private entities in the process decreased progressively as the offer of infrastructural services concentrated increasingly on state-owned monopoly administrations, and as a consequence the state took more responsibility. It was taken for granted that safety standards would be enforced by law. Standardisation in almost entirely state-led committees became an archetype finally. The post-war development of the European Community illustrates the vital importance of the contemporary social environment for the enforcement of standards. Until well into the 1980s, the European countries tried to harmonise a growing number of quality and safety standards through the European Commission as a 'governmental' committee, i.e. by means of directives and regulations. Since the internal market programme of the 1980s, this archetype was changed insofar as only standards of particular importance were set by Community law. The process of harmonising the details of less important standards was 'outsourced' to private standards organisations. In this respect, a mixed private-public standardisation emerged as a new – and old – archetype.

*For traditional infrastructures such as the postal service, that were organised as state monopolies in the nineteenth century, standardisation took place in predominantly state-led committees from a very early stage. Modern infrastructures such as railway traffic that only emerged in the course of the nineteenth century were initially integrated by competition in the market.*

This hypothesis refers to all types of standards, but differentiates between sectors; therefore it follows a synchronic-sectoral approach. By contrast, the last hypothesis referred to a specific type of standard, included all infrastructure sectors and had a diachronic, epoch-specific approach. From the outset – note that the period under study is limited to the nineteenth and twentieth centuries – traditional infrastructures in national markets were characterised by very little or no competition at all. Therefore the parties involved in the international integration of networks were predominantly state administrations which enjoyed a monopoly in their own country. On an international level, there was no competitive market for postal services, as national postal administrations did not attempt to enter foreign markets. Under those circumstances, integration through a committee can be considered as an archetype. However, it was still quite a liberal type as the committee was open to all interested parties and anyone could adopt the standards, but they were not compulsory. For the infrastructures which had only recently emerged at the time like railways, this was different. The market continued to play an important role for quite some time. On the domestic markets, private and public companies competed with each other. The technology was still relatively new. The networks were not closely linked across borders yet. Bilateral and multilateral agreements were made only gradually. Although committees were established at a very early stage in this infrastructural sector, competitive integration played a more important role than in the traditional sectors. The more set the monopolistic structures, the more complicated technology, and the more closely interlinked networks became, the more the profile of the committees rose. In this context a mixed competitive-cooperative type of integration can be identified.

It would evidently be easy to find other examples to establish similar archetypes. All archetypes are 'ideal types' insofar as the complex historical context is whittled down and simplified considerably. Examples not covered by the archetypes mentioned above can easily be found.

The archetypes and hypotheses identified in this chapter are not contradictory, but they overlap and to some extent may be difficult to reconcile. It is important to avoid the impression that historical research has already provided clearly defined archetypes of infrastructural integration or standardisation. So far this has not been the case. For each individual case it is therefore essential to show how types of standards, forms of enforcement and the technical, economic or political environment interact in a particular infrastructural sector. Archetypes in the proper meaning of the word are independent of geography, time and sector and can only be derived from systematic and theoretical contexts. In this respect, the archetypes described in this chapter are located somewhere in between a theoretical construct and an individual historical case.

The examples mentioned show that different perspectives will open up and different levels of analysis will be applied, depending on the chosen starting point – infrastructural sector, type of standard or form of enforcement.

## References

Abbott, K.W., and Snidal, D. 2001. International 'standards' and international governance. *Journal of European Public Policy*, 8, 345–70.

Ambrosius, G. 2009. Standards und Standardisierungen in der Perspektive des Historikers – vornehmlich im Hinblick auf netzgebundene Infrastrukturen, in *Integration und Standardisierung europäischer Infrastruktur in historischer Perspektive*, edited by G. Ambrosius et al., Baden-Baden: Nomos, S. 15–36.

Ambrosius, G, Henrich-Francke, C. and Neutsch, C. (eds) 2010. *Internationale Politik und Integration europäischer Infrastrukturen in Geschichte und Gegenwart*, Baden-Baden: Nomos.

Arthur, B.W. 1989. Competing Technologies, Increasing Returns, and Lock-in by Historical Events. *The Economic Journal*, 99, 116–31.

Auger, J.-F., Bouma, J.J. and Künneke, R. (eds) 2009, *Internationalization of Infrastructures*, Delft: Economics of Infrastructures.

Badenoch, A. and Fickers, A. (eds) 2010. *Materializing Europe. Transnational Infrastructures and the Project of Europe*, Houndmills: Palgrave Macmillan.

Besen, S.M. and Farell, J. 1994. Choosing How to Compete: Strategies and Tactics in Standardization. *Journal of Economic Perspectives*, 8, 117–31.

Blankart, C.B., and Knieps, G. 1992. Netzwerkökonomik. *Jahrbuch für Neue Politische Ökonomie*, 11, 73–87.

Blankart, C.B., and Knieps, G. 1993. State and Standards. *Public Choice*, 77, 39–52.

Blind, K. 2002. Driving Forces for Standardization in Standards Development Organizations. *Applied Economics*, 34, 1985–98.

Blind, K. 2004. *The Economics of Standards – Theory, Evidence, Policy*, London: Edward Elgar Publishing.

Chiao, B. and Tirole, J. 2007. The Rules of Standard Setting Organizations: an Emperical Analysis. *RAND Journal of Economics*, 38 (4), 905–30.

De Vries, H.J. 1998. The classification of Standards. *Knowledge Organization*, 25, 79–89.

De Vries, H.J. 2006a. Fundamentals of Standards and Standardization, in *Standardisation in Companies and Markets*, edited by W. Hesser et al. Hamburg: Verlag Helmut-Schmidt-Universität, 1–33.

De Vries, H.J. 2006b. IT Standards Typology, in *Advanced Topics in Information Technology Standards and Standardization Research*, edited by K. Jakobs, Hershey: Idea Group Inc., 1–26.

Egyedi, T.M. 2003. Consortium Problem Redefined: Negotiating 'Democracy' in the Actor Network on Standardization. *International Journal of IT Standards and Standardization Research*, 1, 22–38.

Farell, J. and Saloner, G. 1985. Standardization compatibility, and innovation. *RAND Journal of Economics*, 16, 70–83.

Farell, J. and Saloner, G. 1988. Coordination through Committees and Markets. *RAND Journal of Economics*, 19, 235–52.

Fischer, R. and Serra, P. 2000. Standards and Protection. *Journal of International Economics*, 52, 377–400.

Gabel, H.L. 2005. *Competitive Strategies for Product Standards*, London: McGraw-Hill.

Genschel, P. 1995. *Standards in der Informationstechnik .Institutioneller Wandel in der internationalen Standardisierung*, Frankfurt am Main, New York: Campus.

Glanz, A. 1993. Ökonomie *von Standards. Wettbewerbsaspekte von Kompatibilitäts-Standards dargestellt am Beispiel der Computerindustrie*, Frankfurt am Main: Lang.

Hofmann, U. 2001. *Netzwerk-Ökonomie*, Heidelberg.

Jakobs, K. 2000. *Standardisation Process in IT: Impact, Problems and Benefits of User Participation*, Braunschweig et al.

Kasper, W. and Streit, M.E. 1998, *Institutional Economics. Social Order and Public Policy*, Cheltenham: Edward Elgar.

Katz, M.L. and Shapiro, C. 1985. Network Externalities, Competition, and Combatibility. *American Economic Review*, 75, 424–40.

Katz, M.L. and Shapiro, C. 1986. Technology adoption in the presence of network externatilities. *Journal of Industrial Economics*, 40, 55–84.

Katz, M.L. and Shapiro, C. 1992. Product Competition and Network Effects. *Journal of Economic Perspectives*, 8 (2), 93–115.

Knorr, H. 1993. Ökonomische *Probleme von Kompatibilitätsstandards: Eine Effizienzanalyse unter besonderer Berücksichtigung des Telekommunikationsbereichs*, Baden-Baden: Nomos.

Mattli, W. 2001. The politics and economics of international institutional standards setting: an introduction. *Journal of European Policy*, 8, 328–44.

Mattli, W. and Büthe, T. 2003. Setting International Standards. Technological Rationality or Primacy of Power? *World Politics*, 56, 1–42.

Molle, W. 1991.*The Economics of European Integration. Theory, Practice, Policy*, Aldershot: Ashgate Publishing.

Nicklas, M. 2000. *Wettbewerb, Standardisierung und Regulierung beim digitalen Fernsehen*, Berlin: VWF/Verlag für Wissenschaft und Forschung.

Schmidt, S. and Werle, K.R. 1998. *Coordinating Technology – Studies on the International Standardization of Telecommunication*, Cambridge, MA, London: MIT Press.

Spruyt, H. 2001. The supply and demand of governance in standard-setting: insights from the past. *Journal of European Public Policy*, 8, 371–91.

Swann, G.M.P. 2000. The Economics of Standardization, in *Final Report for Standards and Technical Regulations*, edited by Directorate Department of Trade and Industry, Manchester: Manchester Business School, 157–92.

Thum, M. 1995. *Netzwerkeffekte, Standardisierung und staatlicher Regulierungsbedarf*, Tübingen: Mohr&Siebeck.

Van der Vleuten, E. and Kaijser, A. (eds) 2006. *Networking Europe. Transnational Infrastructures and the Shaping of Europe 1850–2000*, Sagamore Beach, MA: Science History Publishing.

Voelzkow, H. 1996. *Private Regulierungen in der Techniksteuerung – Eine sozialwissenschaftliche Analyse der technischen Normung*, Frankfurt am Main: Campus.

Weitzel, T., et al. 2000. *Reconsidering Network Effect Theory* , in Proceedings of the 8th European Conference on Information Systems (ELIS 2000), http://www.uni-frankfurt.de/

Werle, R. 1997. Technische Standardisierung im deregulierenden Europa. *Jahrbuch für Neue Politische* Ökonomie, 16, 54–80.

Werle, R. 2001. Institutional Aspects of Standardization – jurisdictional conflicts and the choice standardization organisations. *Journal of European Public Policy*, 8, 393–410.

# PART II
## Case Studies

# Chapter 5

# Two Hundred Years of Failed Tariff Policy: A Comparison of the Reich Freight Tariff Law[1] and the Common Customs Tariff of the EEC[2]

Christian Henrich-Franke

## 1. Introduction

With the German Reich and the European Economic Community (EEC), founded in 1871 and 1957 respectively, two new political systems with similar institutional structures were established (Henrich-Franke 2012). The division of competency and responsibility among the community and its members are comparable in both cases. This is especially true in regard to transport. In both the EEC and the German Reich, a comprehensive transport policy was to be established on the basis of vague and ambivalent regulative clauses.[3] The main element of these transport policies was to be common tariff policy. Although long-lasting and intensive political negotiations were conducted in both cases, both the government of the German Reich and the EEC Commission had to concede, in 1879 and 1972 respectively, that they had failed in respect to these ambitions.

Based on these observations, this report aims to show that the failures of tariff policy have been caused by similar factors. This will also raise questions concerning the value of such a comparison of two different eras.

Here, tariff policy is singled out of the complex field of transport policy. This is due to its exposed position within transport policy topics. Negotiations of tariff policy, however, run parallel to other transport policy measures (e.g. of regulation and capacity), which cannot be discussed here in further detail.

In the next section the various parameters for the comparison of both cases will be presented. These will be parameters for the structuring of presentation and parameters for the analysis, i.e. different factors of failure of tariff policy. Then, the two cases – the Reich Freight Tariff Law and the Forked Tariffs Law of the EEC –

---

1 Reichsgütertarifgesetz.
2 EWG-Margentarifverordnung.
3 From now on, when a direct comparison is being made, the term 'treaty' will be used in reference to both the Reich Constitution and the EEC Treaty.

will be discussed, following a short introduction of the decision-making structures of the regulatory bodies of each. The factors of failure of the tariff policies, which will by then have been differentiated, will be applied for explanatory purposes. In conclusion, a comparison will be made and the method reflected.

## 2. Comparison Parameters

(1) Structuring parameters: In order to make the two cases accessible for a comparative analysis, a simplified version of the policy cycle, which is a sequence model for political processes, will be applied as a presentation structuring tool. This model allows the developments of the tariff policy decision-making processes to be compared with one another. The political decision-making process is divided into several phases, thereby structuring it for analytical purposes (Jann/Wegerich 2003). A simple basic model will be applied, which divides the political decision-making process into an initiative phase, a negotiation phase and a decision-making phase. The *initiative phase* begins with the initial idea and ends when the negotiation of concrete proposals begins. The *negotiation phase* begins with the official proposal submission and ends when the proposal is put to a final vote. The *decision-making phase* begins with the start of the voting session and ends when a final decision is made.

(2) Analysis parameters: For the comparative analysis of the causes for failure, six individual factors have been differentiated and will be repeatedly referred to, using *keywords*, throughout the report. It is important to mention that, with intertemporal comparison, the analysis is limited to political structures and processes as causes for the failure of tariff policy. The tariff policies differ so much in content that this aspect cannot be the basis for a comparison. Transport policies of the nineteenth and twentieth centuries differ significantly in content. Nineteenth-century transport policy was not yet determined by sophisticated theoretical-conceptual considerations, as was transport policy in the twentieth century. To that time, no independent discipline of transport science had been developed, which could have offered such concepts.

The following factors have been differentiated on an empirical basis. They represent neither a theoretical-systematic differentiation, nor do they follow a methodically stringent categorisation or hierarchy. Therefore, the relative importance of the individual factors for the failure of the regulatory processes varies considerably.

*Veto power*: In both the Reich Council (Bundesrat) and the EEC Council, the Federal and Member States, respectively, each had the right to one veto. However, a fundamental difference can be observed between two cases in respect to the veto power. In the EEC Council, each Member State individually had the right to veto. In the Reich Council, the members could only collectively (i.e. with a majority vote) exercise their right to veto. However, an enduring and stable coalition among the Middle powers (Baden, Bavaria, Saxony and Württemberg) in the Reich Council allows us to consider this a functionally-equivalent veto power.

*Separation of initiative and decision-making bodies*: The initiative bodies (Reich Chancellery and EEC Commission) and the decision-making bodies (Reich Council and EEC Council) were clearly separated, while the former was of communal nature and the latter of particular nature. The result was that communal interests were more strongly represented in the initiative phases. In the decision-making phase, particular evaluation criteria were linked to joint consensus. Therefore, the vulnerability of regulatory processes increased in the decision-making phase.

*Vague constitutional terms*: The terms fixed in the treaties (Reich Constitution and EEC Treaty) in respect to tariff policy were formulated only vaguely and did not offer clear objectives. The fact that this lack of focus possessed a structural component complicated matters. These sections laid the foundation for unfocused voting and decision-making structures.

*Negotiation deals/packages*: The willingness to agree on larger, comprehensive deals was lacking among the actors involved in the negotiation processes. Consequent efforts to find compromises between contrasting ideas were made *neither* across different policy areas nor within individual policy areas concerned with different aspects of transport policy.

*Unrealistic expectations of actors throughout the initiative phase*: The goals of the initiators of regulatory processes were often characterised by unrealistic expectations. Throughout the pre-contract stage, interpretations of the proposed policies aimed for goals which, due to the objectively limited political possibilities, proved to be utopian.

*Defence of constituting powers*: The individual political entities attempted to defend their formal constituting powers and to concentrate new powers on themselves. Regulatory proposals were blocked, not due to difference of opinion, but rather due to structural or procedural factors.

### 3. Tariff Policy in the German Reich

The regulatory process in the German Reich can be characterised as a dual process. First, the Reich Chancellery (Reichskanzleramt) or the government of a Federal State could initiate a regulatory development process by presenting a draft proposal to the Reich Council (Bundesrat) for negotiation and approval. In the case of a positive outcome, the Reich Parliament (Reichstag) continued with the assessment and decision-making process. Alternatively, the Reich Parliament itself could initiate a regulatory development process and continue internally with the negotiation process. With a positive outcome in this case, the Reich Council was called on for further negotiation and approval. In both cases, a positive approval vote of both entities was required for the acceptance of a regulatory initiative.

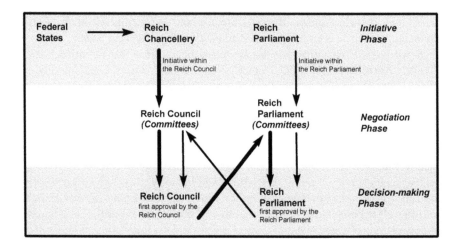

**Figure 5.1    Political decision-making process in the German Reich**

*3.1 Historical Context*

With the establishment of the German Reich in 1871, following the Franco-German War, independent kingdoms and princedoms were united. From then on, these had the status of Federal States. Before 1871, each operated politically independent, at least formally, and was equipped with its own regulatory bodies. The laws regarding railways, especially, were different from state to state. Railways were authorised, planned and regulated independently. Many states even had their own state railways. Throughout the course of the establishment of the German Reich, the new Reich Constitution subjected the railways to common regulations. This gave the Reich the comprehensive right of supervision and control of railway and tariff policy, which could take various forms. Whether the regulation and supervision should fall into the hands of a federal entity, or to the states themselves, was not determined more closely.

The institutions of the newly-established German Reich, in accordance with the decrees of the Constitution, repeatedly attempted to address the alignment of the tariff systems. However, they did not make considerable progress. In 1875, a tariff conference was held, which should have led to the development of a common tariff system. However, due to diverse conflict areas, a proposal could not be agreed upon. In May of 1876, the Imperial Railway Office presented the Reich Council with yet another memorandum of the results of the conference. Still, the Reich Council did not address the question of tariff reform. The Federal States were not especially interested in granting the Reich more influence on the setting of tariffs, as their own budgets depended largely on the tariffs (Moche 1925).

In anticipation of renewed political attempts, the railway management took on tariff policy itself, agreeing upon the so-called 'Reform Tariff' (Reformtarif) in the Spring of 1877. This resulted in a uniform classification of tariffs, but not of uniform tariffs.[4] The Reform Tariff represented a mixture of the value system and carriage system (Wagenraumsystem). As a result, the Reform Tariff designated the following: 2 piece-cargo classes (express and regular) and 4 main classes for wagonload freight with 3 special fares for finished goods, semi-finished goods and raw materials. In regard to freight rates, the Reform Tariff limited itself to the designation of rate limits. Railways continued to have the freedom to implement their own tariff policies for everything below the maximum limits. The railway management still expected a long-term alignment of freight rates (Burmeister 1899). In the short-term, however, this could not be realised. The Imperial Railway Office calculated no less than 63 local, 183 union and 351 special fares for individual items being transported nation-wide. In addition, 199 fares with 314 special fares were counted for individual items being transported internationally (Ulrich 1886).

The heterogeneity of the tariff system was enough reason for a renewed attempt to initiate the development of a tariff law. The final impulse was the transition to a protectionist trade and customs policy in December of 1878 (Torp 2005, 156–83). The Reich, considering the rising expenses and stagnating revenues, needed to be financially independent of the financial contributions (Matrikularbeiträgen) of the Federal States. On account of the unfavourable expense/revenue relationship, the Reich debt increased from 16 million to 139 million Marks between 1877 and 1879. The aftermath of the crisis of the founding years (Gründerkrise) made matters worse. The economy was pleading for protective tariffs. The customs tariff reform focused once more on the goods tariffs. Railway tariffs had similar effects as did customs tariffs. Therefore, tariff reductions in international transport could counteract the effects of the protective tariffs. In order to prevent infiltration of the customs policy and to develop tariff uniformity, the Reich Chancellery initiated the preparations for a Freight Tariff Law.

## 3.2 Initiative Phase

The State governments with railway ownership responded negatively to the idea of the Freight Tariff Law when suggested informally by the Reich Chancellor. However, on February 7, 1879, Bismarck proposed the appointment of a special committee of government experts[5] to develop a Reich law for freight tariffs.[6] It was expected that a special committee of railway experts would remain objective in regard to the contentious issues – supervision and control rights, sovereignty

---

4   Record of proceedings of the general assembly of the Union of German Railway Administrators, February 12–13, 1877, R4101/387, German Federal Archives, Berlin.

5   Printed documents on the Reich Council proceedings, Session 1878/79, No. 27.

6   He was compelled by the same motives described above in the context of the Reich Railway Law (Reichseisenbahngesetz).

and State finances – which had caused the tariff conference to fail. The plan was to call upon the Middle powers to fulfil their duties, rather than merely positioning them before the question of acceptance or refusal of the proposal. The Reich Chancellor speculated that in such a committee, the Reich could obtain a more heavily-weighted voice than it would in a normal committee of the Council.

In order to accelerate the process, the Reich Chancellor invited experts of the various State governments to an informal conference on March 7, 1879, in Berlin. Even before informal talks began, the Middle powers had already agreed upon adopting a negative position.[7] The idea of uniform tariffs was most criticised.[8] The Middle powers believed a flexible tariff policy to be crucial, on account of the regional economic heterogeneity, to reduce the negative effects of economic development. Many State governments regarded it to be too early to judge the effects of the Reform Tariff. Mainly, they did not want to transfer their power to set tariffs to a Reich institution. For them, this step would have meant the loss of sovereignty and a threat to State budgets (*Defence of constituting powers*). It was already clear, at the end of the informal conference, on March 11, 1879, that a Reich Freight Tariff Law would not be passed with the votes of the Middle powers *unless* it was limited to the absolute minimum requirements for the abolition of differential tariffs.

Despite the concerns and fears of the Middle powers, the Reich Chancellor remained firm (*Unrealistic expectations*). He was not willing to sacrifice any of his goals, such as the abolishment of differential tariffs to the benefit of foreign products (*Negotiation deals*). He remained firm in his demands for uniform tariffs, non-discrimination and the appointment of competent Reich institutions for the implementation of tariff policy. Therefore, on March 18, 1879, he called on the Reich Council to finally reach a decision on his previous proposal for the appointment of a special committee (Von der Leyen 1914).

Procedural questions were central to the discussions which followed in the plenary assembly of the Reich Council (Bundesratsplenum). How should the proposed committee be put together and anchored institutionally to the Reich Council? Everybody involved was aware that the location chosen for the negotiations would set the course of the discussions. The informal preliminary discussions, at least, had been characterised by two opposing blocks – (1) the Reich and Prussia and (2) the Middle powers – which confronted one another with very different ideas and goals. A majority vote for one of the two blocks could quickly lead to a pre-emptive determination of the focus of a Freight Tariff Law.[9]

---

7  Statement of the Bavarian Minister of State, Pfretzschner, to the envoy in Berlin Rudhart, 17.2.1879, 5941, Transport Archive, State Archives, Munich.

8  Record of the proceedings of informal counseling of the government representatives at the Imperial Railway Office 7.–11.3.1879, R43/97, German Federal Archives, Berlin.

9  Report of the Württemberg Minister of State Mittnacht to the Baden Ministry of State on the Reich Council session on March 29, 1879, March, 30, 1879, Bü 3594, E130, State Archives, Stuttgart.

Contrary to Bismarck's vision, the Reich Council proposed, on April 4, 1879, a committee of tariff experts as a new Freight Tariff Committee of the Reich Council (Gütertarifausschuss des Bundesrats), as proposed by Württemberg. In this, the Middle powers were successful in establishing a committee within which they would have a large percentage of the vote (*Defence of constituting powers*). This committee was to develop a draft of regulation, which would then be formally taken over by Reich Council as a regulatory initiative (Reichert 1962, 75). The committee, in preparation, called upon the Imperial Railway Office to prepare a draft of the discussion foundations. In doing so, they also took over the content of the actual initiative. This demonstrates that the Middle powers had little real interest in the content of the regulations. Primarily, they wanted to maintain control over the development process.

The Imperial Railway Office (which was closely aligned with the Reich Chancellor) responded quickly and presented a result on April 28, 1879 (Kunz 1996, 5–26). Revealingly, the provisions of the initial regulatory draft about freight tariffs ignored almost entirely the concerns voiced in the informal preliminary discussions (*Separation of initiative and decision-making bodies*). Although Saxony had doubted the need for freight tariff regulation and emphasised the positive developmental tendencies of the Reform Tariff, the drafted proposal went beyond everything mentioned up to this point and included even the passenger fares. Although the Middle powers had marked tariff uniformity as being an obstacle for the heterogeneous regional economies, the draft proposed general tariff uniformity with few exceptions. Beyond this, the tariffs were to be determined by Reich institutions (Reich Council and Imperial Railway Office), which would have made difficult a regional tariff policy. Finally, the draft proposed an obvious extension of Reich competency, which had been rejected by the State governments in railway laws and the tariff conference of 1875, and which was not covered by the terms of the Constitution[10] (*Vague constitutional terms*).

### 3.3 Negotiation Phase

The State governments viewed this draft as a formal regulatory initiative, rather than a foundation for discussion, and entered into the negotiations (Rottsahl 1936). On May 6, 1879, only two days after forwarding the draft to the State governments, the Middle powers entered into coordinating talks about the content of the proposal. By the end of April, permanent contact was kept to the envoys in order to develop common negotiating strategies.[11]

The specially-constituted Freight Tariff Committee of the Reich Council entered into negotiation on the basis of the initial draft proposal on May 9, 1879.

---

10 Draft of regulation regarding tariffs of German railways, April 28, 1879, R43/97, German Federal Archives, Berlin.

11 Correspondence of the Württemberg State Ministries in regard to the Freight Tariff Law, Bü 3594, E130, State Archives, Stuttgart.

The governments were passive in these discussions – as was normal for special Reich Council committees – and did not interfere with the negotiation process. The experts, which were recruited primarily from among the general directors of the railways, were left with the most negotiating power in regard to the development of tariffs and appropriate tariff structures. As a result of the congruent interests of the governments and the State railway managers, there was little opportunity for intervention. Additionally, the experts knew very well that their governments would not concede in regard to their own tariff policy notions.[12]

Contrary to the initially disadvantageous situation, the Reich Council committee was successful in negotiating a common proposal. This was possible, in part, because the same individuals were negotiating with one another as had negotiated the tariff reform of the Union of German Railway Administrators (VDEV). Furthermore, the negotiations of the Reich Council committee had no binding effects on the later vote in the plenary assembly of the Reich Council. The proposal was weakened considerably because decisions on three key aspects – determination of uniform tariffs, power to determine tariff exemptions and regulation of tariffs over distance – could only be made on the basis of a controversial vote. In the Reich Council committee, a simple majority was enough. The Reich representatives outvoted the representatives of the Middle powers with changing coalitions (Reichert 1962, 76). All powers, regarding the named aspects, were to be transferred to the Reich Council. Thereby, the proposal intended to transfer the central tariff policy competencies of the States to the Reich, thereby making independent tariff policies impossible. The approval of the States in the plenary assembly of the Reich Council seemed unlikely (*Defence of constituting powers; unrealistic expectations*). At the beginning of May, before the Reich Council committee met for the first time, the governments of the Middle powers met amongst themselves to decide whether or not they should reject a regulatory proposal by bringing the argument of a 'proposed changing of the terms of the Constitution'[13] (*Vague constitutional terms*).

### 3.4 Decision-making Phase

When the proposal[14] of the Freight Tariff Committee was presented to the Reich Council on June 6, 1879, it met with considerable resistance. The representatives of the Middle powers contested, once again, the controversial points of the law. They criticised the transfer of the power to set tariffs to the Reich and cautioned that the Reich Constitution spoke only of a Reich power to regulate tariffs.

---

12   Report of the General Director of the Royal Bavarian Transport Institutions Hocheder, the Bavarian representative to the Freight Tariff Committee, to the Bavarian State Ministry, May 17, 1879, 5941, Transport Archive, Bavarian State Archives, Munich.

13   Report on a meeting of the Bavarian envoy with the Württemberg Minister of State Mittnacht, May 6, 1879, 5941, Transport Archive, Bavarian State Archives, Munich.

14   Printed material on the negotiations of the Reich Council, Session 1878/79, No. 101.

The transfer of the power to set the tariffs, according to the Middle powers, amounted to a change of the terms of the Constitution. They wanted to make clear to the Reich leadership that they would support the underpinning of the customs policy through an appropriate Freight Tariff Law, but not the transfer of power to calculate tariffs and uniform rates to Reich institutions. If the Reich and Prussia were unable to accept these conditions, the 'constitutional question' would be posed (*Defence of constituting powers*).

The negotiations were stuck in a difficult position. If the law was to be interpreted as a change of constitutional policy, the Middle powers would have a comfortable blocking minority in the Reich Council (*Vague constitutional terms, Veto power*). If, however, the law was not to be interpreted as a change of constitutional policy, then a simple majority would be sufficient to pass the proposal. So, if a decision was to be made, it could only be a political one. Every attempt to outvote the Middle powers in the Reich Council would be, in this case, legally controversial.

Regardless, the proposal of the Freight Tariff Committee was brought to a vote. It maintained the absolute majority, but not the votes necessary for a change of constitutional policy. This sparked an extended legal dispute on whether the proposal had now been passed or blocked. Saxony, Baden and Württemberg, convinced that their position was correct, forwarded the proposal to the Constitutional Committee of the Reich Council (Verfassungsausschuss des Bundesrats) for final clarification.

It did not come, however, to a final clarification on the constitutional matter. The Württemberg Minister of State, Mittnacht, agreed to a compromise with Reich Chancellor Bismarck on June 21, 1879. For the Reich and Prussia, as well as for the Middle powers, the easiest solution seemed to be to ignore the constitutional question and tolerate the Freight Tariff Law for an undefined period of time. Saxony, Baden and Württemberg withdrew their request for a constitutional assessment. In return, the proposal was forwarded back to the Freight Tariff Committee to '*gain time to come to a material understanding before the next session*'.[15] Here, however, the proposal for the Freight Tariff Law was never put to discussion.

## 4. Tariff Policy of the European Economic Community

The regulatory process of the EEC can be described as a direct path with additional, optional iterative processes. The Commission initiated each regulatory process with a draft proposal. Each proposal was forwarded to the Commission Presidency, which then had to obtain obligatory statements from the Economic and Social Committee and the European Parliament. Then, the Commission could revise the proposal, before putting it to negotiation before the Council, the Committee of

---

15    Statement of the Württemberg Minister of State Mittnacht to the Bavarian Minister of State Pfretzschner, May 28, 1879, 9244, Ministry of Foreign Affairs, Bavarian State Archives, Munich.

Permanent Representatives and the Transport Group. Finally, the Council voted on the proposal, with three possible outcomes: approval, request of a revised proposal or rejection.

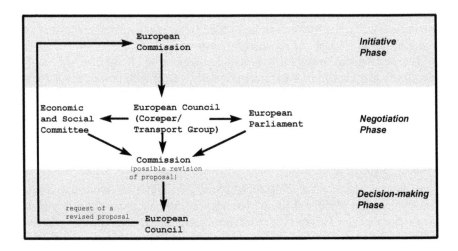

**Figure 5.2     Political decision-making process in the European Economic Community**

*4.1 Historical Context*

As the EEC Commission, in the context of drafting a comprehensive European transport policy, turned to the aspect of tariff policy, it was confronted with a multi-faceted problem complex. In addition to the interdependencies among capacity, tariff and competition policy, the individual tariff systems differed from another in regard to Member State, transport carrier, international transport relations and methods, and pricing policy. Multilateral European tariff treaties and corresponding experiences, on which the Commission could have based its efforts, were hardly at hand. In addition, the transport section of the EEC Treaty contained few material provisions for a common transport policy (Dumoulin 2007, 475–88).

Following lengthy preparations, the Commission presented a memorandum in April of 1961, which was followed by an Action Programme in May of 1962. The content of the Action Programme amounted to an extensive interpretation of the EEC Treaty. The Commission made clear that the common transport policy should not be limited to minimal measures for the dismantling of interferences and distortions in freight transport. The Commission made an effort to suggest a pragmatic approach of parallel measures of deregulation and alignment.[16] Between ideal-typical notions of the deregulation of an economic sector and a specially-regulated economic sector, compromise was difficult to come to (Ebert/ Harter 2010, 63–72).

In regard to tariff policy, the Action Programme for Forked Tariffs suggested a compromise of a fixed-price system and a free-pricing system.[17] The Commission recognised positive pre-conditions for establishing competition within and among transport carriers on the basis of cost-oriented tariffs. Due to the existing differences, the tariff policy measures were to be carefully introduced. Passenger transport and occasional transport would be excluded and the focus set exclusively on freight transport.[18] However, it must be mentioned that abstract models played a greater part in the Action Programme than did practical suggestions. In respect to concrete implementation, the Action Programme offered little.

The discussions of the Action Programme had not brought the Council any closer. Despite this, the Council formally requested, on March 8, 1963, that the Commission should present concrete regulatory drafts with the aim to realise a common transport policy. The Council withdrew from the negotiation and decision-making arenas and left the compromise process to the Commission and its Directorate-General for Transport. Only the Dutch had threatened to veto in the Council. The Dutch wanted navigation of the Rhine to remain independent of EEC transport policy, while also disagreeing with the approach of the tariff policy. They benefitted especially from exemptions on the Rhine, which were guaranteed by the Mannheim Act, and transported the majority of Rhine shipments from the harbour at Rotterdam. As their own system of transport was the most cost-efficient in Europe, they viewed their liberal approach to transport policy as being correct.

*4.2 Initiative Phase*

On May 21, 1963, the Commission presented a regulatory draft of an obligatory forked tariff system for all three inland transport carriers (Heinrich-Franke 2012). The Commission had made fundamental decisions on several controversial points. Due to the previous discussions of the Council, these had little chance of being realised (*Unrealistic expectations*). The draft declared only shipments weighing less than 5 tonnes, transport distances of less than 50km and long-term rental agreements as being exempt from the forked tariffs. In doing so, the Commission planned to include navigation of the Rhine in the common tariff policy, despite the threats of the Dutch. It is not certain whether or not it was strategically calculated that the discontinuation of the veto power in the Council, on January 1, 1967, would

---

16   Sixth General Report on the Activities of the Community, p. 192.
17   Third General Report on the Activities of the Community, p. 202.
18   Sixth General Report on the Activities of the Community, pp. 188–95 (short version); VII/KOM(62)88 (full version).

force the Dutch to yield. It is certain, however, that the Commission was not willing to budge on its independent initiative power (*Defence of constituting powers*).

For other open questions, the Commission made decisions in the draft which the Council had not been able to make. (1) In regard to the scope of the forked tariff system, validity was to be extended to inland and international freight transport. (2) The regulation provided for a range of between -10 percent and +30 percent of the price upper limit for the free-pricing system. The Commission, which wanted to set prices itself, planned to consider the market situation in addition to technical and socioeconomic developments. (3) The Commission struggled on the question of obligatory publication of margin limits. These were to be published alongside actual prices. Details in this regard, however, were not described in the draft (Ebert/Harter 2010, 91–3) (*Unrealistic expectations; separation of initiative and decision-making bodies*).

In order to alleviate agreement for the Member States, the draft proposed long transition periods, extensive differentiation and diverse national freedoms for the implementation of a forked tariff system for inland transport. Nevertheless, the Commission failed to consequently link the forked tariff draft to other transport policy proposals. The Commission kept the national interests within the individual regulatory proposals, but did not extend them further. Thereby, the basis for compromise was reduced (*Negotiation deals*).

*4.3 Negotiation Phase*

The regulatory draft of the forked tariff system did not offer the compromise solution which the Council had hoped for. The Dutch, in keeping with their previous position, rejected the forked tariff system. The Dutch argued that forked tariffs could only counteract the symptoms, and not the causes, of disturbances in transport markets. Furthermore, they gave priority to capacity policy and the exemption of navigation of the Rhine.[19] It would not be fair, however, to interpret the position of the Dutch as an overall rejection of a common transport policy. As long as the transport system of the Netherlands was the most cost-efficient system in the EEC, it viewed its own liberal transport policy as being the most promotable approach. In any case, navigation of the Rhine would be appropriately regulated by the Central Commission for the Navigation of the Rhine. The EEC Treaty had remained vague and interpretable in regard to navigation of the Rhine, just as it had on many points (Heinrich-Franke/Tölle 2011, 331–52) (*Vague constitutional terms*).

While the Dutch cemented their position in the Council negotiations, the remaining five EEC Members continued to align with one another. The idea of a four-year trial phase of the forked tariff system, for all three transport carriers of

---

19   Report of the Dutch representatives in Brussels on the discussions of the Transport Group on July 29–30, 1964, MBZ(55-64)-99-3484, Archief van Ministerie van Buitenlandse Zaken, Den Haag.

international cargo with extensive diversification of tariffs, was much supported. Further negotiation was necessary in regard to questions of the application of forked tariffs for inland transport, the margin scope, tariff levels and (unpublished) special agreements.[20]

The Transport Group was occupied with the tariff policy suggestion of the Commission for the whole of the Fall of 1964, but was unable to reach an agreement. Remarkably, there were no renewed attempts to negotiate deals/packages within and beyond transport policy (*Negotiation deals*). The positions of the delegations in regard to the simultaneous implementation of the forked tariff system remained unchanged. Germany, France and Luxemburg categorically rejected the exemption of navigation of the Rhine. Thereupon, the Council President had no choice but to sum up the unwavering, contrasting positions on December 10, 1964. Informally, the Council President asked the Commission to come to a compromise. It was to review the regulatory proposal and present a modified solution.[21]

The Commission then tried to assess the national positions and potential retreats in bilateral negotiations with the transport ministries of the Member States, before defining the content of its proposal. The Commission did not plan on backing away from the equal treatment of the transport carriers. Opposite the transport ministries, the Commission firmly stated that it was not questioning the system as a whole, but trying primarily to find a solution to the problem of integrating Rhine navigation into the forked tariff system. Thereby, the Commission remained firm on the controversy surrounding the Dutch (*Unrealistic expectations; separation of initiative and decision-making bodies*).

The Commission presented its results to the Council during its session on March 9, 1965. It held on to a forked tariff system for all three inland transport carriers, including Rhine waterways. Furthermore, it rejected the separation of the Rhine waterway network from the other inland waterway networks. According to the Commission, uniform regulations were necessary due to the fluid transitions between the waterway networks.[22] In return, the Commission proposed a very liberal tariff policy solution for inland waterway navigation and a differentiated forked tariff system for the three internal transport carriers. A liberal tariff system for the railways would not meet consent in the Council. Thereby, the Commission backed away from the ideal of a uniform forked tariff system, which worked in favour of stronger nuances in regard to the transport carriers.

---

20    Internal report of the German Federal Ministry of Transport on the discussions of the Transport Group on July 29–30, 1964, B108/10163, German Federal Archives, Koblenz.

21    Final report of the German Federal Ministry of Transport on the 155th session of the EEC Council of Ministers on December 10, 1964, B108/13114, German Federal Archives, Koblenz.

22    Proposal from the EEC Commission to the Council on March 9, 1965, about the introduction of a forked tariff system for railway, road and inland navigation freight transport, S.10, B108/13113, German Federal Archives, Koblenz.

To enable constructive negotiations in the Council, the Commission did not present a closed concept. Rather, it presented four possible tariff systems for inland navigation and their possible effects on the other transport carriers. As a side note, the Commission also presented a system of reference rates (reference margin system), which had been suggested by the French. According to this system, the administrative bodies responsible for inland navigation would be required to determine (reference) tariffs. These would not be obligatory. Enterprisers were to remain free in determining their prices and be obligated only if their actual prices fell outside of the margins.

Meanwhile, the French government, which had taken over the Council Presidency on January 1, 1965, dared a tariff policy offence, which competed with the regulatory proposal of the Commission. Several reasons had moved the French government to do so: (1) The French government feared unwanted majority decision-making, regarding tariff policy, in the Council following the discontinuation of the veto power rule on January 1, 1967. (2) In the debate on general integration policy and the future of the EEC, the conflict on whether to strengthen supranational or intergovernmental elements had sharpened. If France could free the deadlocked transport policy, it would undoubtedly support the strengthening of intergovernmental elements. The French saw the key to finding a transport policy compromise as being a reference margin tariff system for all of inland navigation. This concept was based on free-pricing, but required the publication of prices once they exceeded a 15 percent margin surrounding an indicative tariff.[23]

After several bilateral talks, the Dutch agreed to the French proposal. The French offer was very accommodating to the wishes of the Dutch. It provided for the special treatment of Rhine navigation as well as for the beginnings of an overall liberalisation process of the other inland transport carriers. Furthermore, the French proposal came at the right moment, as the Dutch also were wary of the discontinuation of the veto power rule. The same calculation led the other Member States to agree in succession. They considered the inclusion of Rhine navigation to be a valuable symbolic success.

In May of 1965, the French proposal was officially presented to the Transport Group and negotiated. The Commission was once again intensively included in the discussions. The Commission could only participate on account of political considerations, but did not want to seem a breaker of the transport policy. It did not show considerable enthusiasm, as it disagreed in equal measure with the process surrounding the development of the proposal as with its content.[24]

---

23   Telex of the Dutch ambassador in Paris, Benting, to the Foreign Ministry, January 29, 1965, MBZ(65-74)-996-3171, Archief van Ministerie van Buitenlandse Zaken, Den Haag.

24   'The representatives of the Commission participated only by observing and answering specific questions, without agreeing to any version straying from that which was

In contrast to the Commission, the French Council President had linked the tariff policy proposal with other policies as a regulatory package.[25] First, depending on the transport carrier, either an obligatory or a reference margin tariff system would be applied. The compromise would not, however, exclude special agreements outside of the margin. Second, the obligatory and reference tariffs would be published by the Member States. To promote market transparency, special institutions, such as freight exchanges and freight offices, would be established. The more highly disputed question of whether tariffs were to be verified initially, or whether a subsequent verification or justification process should occur, remained unanswered. Third, the compromise provided for a 'Market Monitoring Committee' (Ausschuss zur Überwachung des Marktes), put together of government representatives, which would be chaired by representative of the Commission. An understanding had not been achieved in regard to the competency and tasks of the tariff committee (*Constituting powers*).

The compromise proposal for the tariff policy seemed, upon closer inspection, to be an intermediate step. It presented a vague concept, rather than a compromise which had been negotiated in detail. Difficult problems, such as the opening and differentiation of the margins, the relationship to third countries, and the handling of ECSC products, still needed to be solved.[26] The Council had simply suspended decision-making on many controversial details. The French proposal was a compromise between conflicting, particular interests. An open formulation had been chosen, so that the compromise could be interpreted and clarified in a number of ways. The Dutch interpreted it as though the other delegations had accepted a free-pricing rule, while other States, such as the FRG, did not entirely exclude the possibility of administrative intervention in the setting of tariffs.[27]

## 4.4 Decision-making Phase

The Council passed the proposal, which had been reworked by the Transport Group, with minor changes on June 22, 1965 as an 'Agreement on the organisation of the transport sector'. This resonated with the public. This agreement has often been viewed as being the first step to a common transport policy (Hallstein 1969, 221). The compromise still contained many unanswered questions. Therefore, the Committee of Permanent Representatives was called on

---

proposed by the Commission', Weekly report of the German Federal Ministry of Transport, May 17, 1965, B108/13115, German Federal Archives, Koblenz.

25    For the suggestion of the Transport Group: Note on the organisation of the transport market, June 15, 1965, BAC 14/1967/8, Historical Archives of the European Commission, Brussels.

26    Report of Department A2 of the German Ministry of Transport on the state of the EC transport policy, July 1, 1965, B108/13116, German Federal Archives, Koblenz.

27    Report of the German Ministry of Transport on the Council session on June 22, 1965, B108/13115, German Federal Archives, Koblenz.

to decide upon the final system and to develop a suggestion for the next Council session. Under the pressure applied by the French, the Council had strengthened the intergovernmental element and robbed the Commission of its initiative monopoly. This offence of the Council against the actual decision-making process as defined in the EEC Treaty worked at the expense of the Commission.

Just as the tariff policy seemed to be making progress, the 'empty chair crisis' temporarily paralysed the community (Ludlow 2006). France distanced itself from the EEC institutions and, in doing do, blocked the decision-making process of the Council. The tariff policy dynamic came to an abrupt halt, as all decision-making capabilities of the EEC were blocked. When, in January of 1966, the crisis was resolved, the Luxembourg Compromise had extensive consequences for the EEC, in general and specifically for transport policy. From then on, every Member State would have the right to veto in the Council, as long as *'very important'* national interests lay on the line (Brunn 2002, 370). The Luxembourg Compromise created a new reality of transport policy and negotiation tactics (Henrich-Franke 2009, 135–40). The Dutch, especially, who no longer had to fear a qualifying majority in the Council after January 1, 1967, returned to their previous veto stance[28] (*Veto power*).

The political crisis had put the Commission, which was the only fully-functioning EEC institution, in a favourable position. In June of 1965, it energetically began with the drafting of a revised version of the forked tariff system. Due to the discontinuation of the formal decision-making process through the Council, consultation of the Transport Group was expected. On several points, the Commission distanced itself from the Council compromise. Independently, it made several substantial changes in the direction of its regulatory proposal of March of 1965, even though this amounted to a distancing from realistic goals (*Unrealistic expectations*). This approach legitimated the Commission in regard to its initiative monopoly (*Separation of initiative and decision-making bodies*). The Commission, in the draft proposal which was presented in September of 1965, formulated conditions about the cost-coverage of transport services, strengthened the regulations on price-setting and created possibilities for administrative intervention – in the form of set, limited-time forked tariffs – in order to prevent the abuse of market-controlling positions. The failure of the Commission to purposefully integrate the forked tariffs with other transport policy measures was of great importance. The negotiation package, which had been discussed in the Council, was re-opened (*Negotiation deals*). The Commission re-opened all of the old conflicts regarding the content of the policy. It continued to hope, that the discontinuation of the veto power rule would force the Member States

---

28   Report on a meeting of the Director for International Transport Policy, Raben, with colleagues of the DIE, June 1966, MBZ(65-74)-996-3174, Archief van Ministerie van Buitenlandse Zaken, Den Haag.

to agree. The Commission, however, did not and could not have expected the Luxembourg Compromise.[29]

With the combination of the changes of content and the reintroduction of the veto power rule, the agreements were soon forgotten in the Spring of 1966. Following controversial negotiations during the Summer and Fall, the Council found itself once again caught in a deadlock during its session on October 19–20, 1966. A tariff policy agreement was not possible, as the majority of the Member States tried to realise their particular interests (*Separation of initiative and decision-making bodies*). The Ministers saw no other way out than to put aside the tariff policy, which should have been a key element of a common transport policy ever since the initial Commission memoranda of April 1961. A forked tariff system, which would have included all three inland transport carriers, had failed (Frerich-Müller 2004, 13–14).

## 5. Conclusion

It can be concluded that the six different factors introduced at the beginning of this chapter are inherent to both attempts of tariff policy integration. The extent to which transport and tariff policy translates to power politics is evident. The struggle for constituting power turned out to be a central element in both cases. Overall, the involved actors did not yet know how to handle the new political systems. The separation of communal and particular decision-making institutions was not properly used to develop intensive dialogues in the initiative phases, which would have more easily led to compromise. Rather, compromise was prevented by the failure to recognise what was politically realistic and by the permanent use of veto power. The importance of the veto power rule was demonstrated in regard to the forked tariffs. In this case, the Member States proved to be much more willing to compromise if there was a threat of losing their strongest weapon – veto power. It is remarkable to observe that the initiators of regulatory processes in both cases, at least from a historian's perspective, falsely calculated what might have been realistic. For this reason, they did not attempt 'compromise solutions' of negotiation deals and packages. The vague treaty terms also contributed significantly. The vague terms of the Reich Constitution led to the failure of the Reich Freight Tariff Law, even though a decision had already been made in regard to the content.

The comparison of the failed tariff policies of the German Reich and the European Economic Community demonstrates that, in focusing on political structures and processes, certain factors can be filtered out. These factors had similar effects (independent of time and context) on both political systems. If

---

29  Briefing of the Commission to the Council: The common transport policy following the resolution of the Council, October 20, 1966, BAC 3/1974/79, Historical Archives of the European Commission, Brussels.

these factors led, in both cases and despite the different conditions of the ninteenth and twentieth centuries, to similar outcomes, then the similarities gain a special explanatory importance. If one would go so far as to generalise the results of this historical comparison, several theses could be formulated. These theses might offer relevant suggestions for current political situations. The following three theses, for the conception of institutional structures in international relations, can be developed on the basis of the results of this chapter:

- Possible effects of competition between institutions and organisations should be assessed and, if possible, productive competition supported;
- The separation of initiative and decision-making responsibilities among the various decision-making levels (Member States and communal institutions) should be avoided, as this often results in unproductive competition. Instead, a strong integration of these levels in every phase of the decision-making process can be recommended;
- The modification of political structures and, especially, a change in power dynamics should be avoided in the early phases of common policy and regulatory processes.

Whether or not the empirical basis of this chapter is sufficient for drawing general conclusions or for developing a theoretical foundation remains to be seen. In the end, the comparability and the importance of the content of policy must be sought after. It seems that tariff policy, especially, within the complex field of transport policy, is difficult to integrate. Therefore, content cannot be ignored. Comparability, at least on a macro-level, must remain in focus.

### References

Brunn, G. 2002. *Die Europäische Einigung von 1945 bis heute*. Stuttgart: Reclam.
Burmeister, H. 1899. *Geschichtliche Entwicklung des Gütertarifwesens der Eisenbahnen Deutschlands*. Leipzig: Duncker & Humblot.
Dumoulin, M. 2007. Der Verkehr. Bastion nationaler Pfründe, in *Die Europäische Kommission, 1958–1972,* edited by M. Dumoulin et al. Luxembourg: Amt für amtliche Veröffentlichungen der Europäischen Gemeinschaft, 475–88.
Ebert, V. 2010. *Korporatismus zwischen Brüssel und Bonn: die Beteiligung deutscher Unternehmensverbände an der Güterverkehrspolitik (1957–1972)*. Stuttgart: Steiner.
Ebert, V. and Harter, P. 2010. *Europa ohne Fahrplan? Anfänge und Entwicklung der gemeinsamen Verkehrspolitik in der Europäischen Wirtschaftsgemeinschaft (1957–1985)*. Stuttgart: Steiner.
Frerich, J. and Müller, G. 2004. *Europäische Verkehrspolitik. Von den Anfängen bis zur Osterweiterung der Europäischen Union*. München: Oldenbourg.
Hallstein, W. 1969. *Die Europäische Gemeinschaft*. Düsseldorf: Econ Verlag.

Henrich-Franke, C. 2009. Gescheiterte Integration. Die Europäische Wirtschafts-gemeinschaft und die Formulierung der gemeinsamen Verkehrspolitik (1958–1967). *Journal of European Integration History,* 15(2), 127–50.

Henrich-Franke, C. 2012. *Gescheiterte Integration im Vergleich: Der Verkehr – ein Problemsektor gemeinsamer Rechtsetzung im Deutschen Reich (1871–1879) und der Europäischen Wirtschaftsgemeinschaft (1958–1972).* Stuttgart: Steiner.

Henrich-Franke, C. and Tölle I. 2011. Competitions for European competence: the Central Commission for Navigation on the Rhine and the European Economic Community in the 1960s. *History and Technology,* 27(3), 331–52

Jann, W. and Wegrich, K. 2003. Phasenmodelle und Politikprozesse: Der Policy Cycle, in: *Lehrbuch der Politikfeldanalyse,* edited by K. Schubert, and N. Bandelow. München: Oldenbourg, 71–105.

Kunz, W. 1996. Vom Reichseisenbahnamt (1873 bis 1919) zur Gegenwart. *Jahrbuch für Eisenbahngeschichte,* 28(1), 5–26.

Ludlow, P. 2006. *The European Community and the Crises of the 1960s. Negotiating the Gaullist challenge.* London: Routledge.

Moche, H. 1925. *Die deutschen Eisenbahngütertarife in ihrer geschichtlichen Entwicklung.* Köln: Universitätsschrift.

Reichert, H. 1962. *Baden am Bundesrat 1871 bis 1890.* Heidelberg: Universitätsverlag.

Rottsahl, R. 1936. *Bismarcks Reichseisenbahnpolitik.* Gelnhausen: Kalbfleisch.

Torp, C. 2005. *Die Herausforderung der Globalisierung, Wirtschaft und Politik in Deutschland 1860–1914.* Göttingen: Vandenhouk & Ruprecht.

Ulrich, F. 1886. *Das Eisenbahntarifwesen.* Berlin: Guttentag.

Von der Leyen, A. 1914. *Die Eisenbahnpolitik des Fürsten Bismack.* Berlin: Springer.

Chapter 6

# Visions of Rail Development –
# The 'Internationality of Railways' Revisited[1]

Martin Schiefelbusch[2]

## 1. Introduction

Throughout its individual contributions, this book adopts a wide definition of 'integration', taken to comprise all processes which reduce 'friction' in the working of social and technical systems. In the field of infrastructure, the main challenge is to achieve coherent, interoperable networks across geographical boundaries. The 'standard' discussed in this chapter is the *understanding of key stakeholders* in the railway industry regarding the *attitude to be taken towards international cooperation* in the development *of rail passenger services*. This is 'measured' in the prominence of issues relating to international versus domestic market and service development in the industry and the opinion on cross-border activities. We compare the situation in the late nineteenth-early twentieth century with that about 60 years later, hence two periods in which the basic preconditions for international services – a linked-up network – were in place, and ask what has been made out of this asset.

This 'standard' is without doubt a very 'soft' one compared to the other examples in this book. From a 'technical' point of view, this example may seem less challenging and therefore of less interest. But, on the other hand, this situation leads to new questions: Did the different stakeholders – railway providers, government, local authorities, rail users – take the existing level of service provided for granted or not? Who expressed which demands? To what extent was the industry's view shaped by the different national settings? From a political and social science point of view, such 'weak' standards can therefore even be a more interesting object of study, precisely because the need for 'standardisation' is less obvious, leaving more room for the parties involved to develop their position.[3]

---

1  The title of this chapter refers to the conference and book 'Internationalität der Eisenbahn' (Burri 2003).

2  The author gratefully acknowledges the assistance of Melissa Gómez and Sonja Ziener in the synthesising of material and the preparation of this article.

3  Due to space limitations, only selected issues can be presented here in detail. For a comprehensive presentation, readers may refer to Schiefelbusch 2012.

## 2. The Railways and their Specific Integration Challenges

Although railways were first built as individual lines in 'niches' where they were particularly useful or circumstances favourable, their potential as a *means of integration* and the benefits of a comprehensive network were quickly recognised. The idea of Europe as a single market, as a (tourist) arena of experience as well as a space of self-realisation for the civil identity against the still monarchist and aristocratic government with its political and military elites, became visible through the railways and fostered civil self-confidence on a massive scale. Even prior to the construction of the first railway lines in Germany, regarding this technology the vision of a civil European empire of peace was promoted. The well-known German rail visionary Friedrich List predicted 1837:

> How quick will national prejudices, national hatred and national selfishness among the cultivated peoples give room to deeper comprehension and better emotions, if individuals of different nations are linked through thousand ties of science, art, trade and industry, friendship and family relations. (List in von Rotteck 1837)

That was eight years before the first international railway line in Europe was built from Aachen in Prussia to Liege in Belgium.

At the same time railway technology is an exceptionally challenging and complex *object of integration*. An 'integrated' railway traffic, meaning its seamless flow across borders, requires the establishment of joint – at least compatible – standards in many fields. It is not sufficient to use the same track gauge, there have to be as well agreements on rolling stock, operative regulations and signalling, energy supply, documentation and clearing of services, configuration of prices and products and much more. Further challenges arise from the *overlapping of different demands* on the rail network. While local and long-distance transport easily mix on the roads, and national and foreign mail is processed in the same post office, this is not readily the case with the railways: Trains stopping frequently or travelling at low speed for other reasons occupy a bigger part of track capacity than those passing without any stop at line speed. Resulting losses in capacity can be minimised through sophisticated timetable design, but solved only through the construction of costly sidings (Schiefelbusch 2010).

Considering this, the rationale for studying 'visions of international railway development' can be summed up as follows: The benefits of a rail network that provides good 'integration' can easily be recognised in reduced transport time and cost, but they come, on the other hand, at a price: Extra efforts are necessary to negotiate and maintain 'standards' in the interest of integration even in cases where the need for it may not be obvious. As most traffic is usually national rather than cross-border, decision makers might well ask themselves whether coordinating one's own activities with the neighbours is worth these efforts (cf. de Bruin 1971).

## 3. The Case Studies

To get an overview of the debate on 'internationality' with a limited resource input, we limited the material used to the periodicals issued by the main railway industry associations and some archive sources consulted as part of the overall work on this project. For the first period, this was the journal of the Association of German Railway Administrations ('*Zeitung des Vereins deutscher Eisenbahnverwaltungen*')[4] between 1878 and 1914 as well as the limited records of this organisation kept in the German Federal Archive.[5] This journal informed about practical development in the railway industry (new routes, tariff changes etc.) and was used to publish new ideas and discuss them. Coverage of controversial issues like the pros and cons of different fare systems often extended over several issues including initial articles, responses and counter-responses. A disadvantage of this source is that articles were often published anonymously, making it difficult to interpret the author's position. It should be noted that, in spite of its name, the Verein (Association of *German* Railway Administrations) had allowed non-German or Austro-Hungarian companies as full or associated members almost since its foundation (Kaessbohrer 1933), and thus can be considered an international organisation, although the German speaking countries clearly dominated.

For period 2, the following journals were examined: *Die Bundesbahn* (the main professional journal issued by the railways in the Federal Republic of Germany), *Internationales Archiv für Verkehrswesen/Internationales Verkehrswesen, Railway Gazette/Railway Gazette International, Revue Generale des Chemins de Fer* and *Rail International* (from 1970), the latter issued by the International Union of Railways (Union internationale des chemins de fer, UIC). Particular attention was given to reports on international projects as well as to relevant editorials, which were often written by leading railway managers or transport ministers. Archive material for this period was taken from the UIC documentation centre[6] and the German Federal Archive.[7] This material contained information on the development of the Trans Europ Express (TEE) passenger services in the 1950s (cf. section 5.3) and policy documents on the future of international passenger services in the early 1970s. A limitation of this selection

---

4  The abbreviation 'Zeitung' with year and page number is used as reference in the following. Articles are shown with their headline due to the frequent lack of a named author. The short forms 'Verein' or 'VDEV' are used to refer to the organisation.

5  Bundesarchiv Berlin, Section R 4307 (referred to as BA B in the following).

6  UIC conférences commerciales voyageurs, Groupe Permanent Commun: Avenir du Trafic International voyageurs, Sous-commission mixte commercial-mouvement, and Groupe ad hoc Confort Optimum Voyageur (referred to as DOC UIC in the following).

7  Bundesarchiv Koblenz (referred to as BA KO in the following), Sections B 108 (Federal Department for Transport, Bundesministerium für Verkehr, BMV) and B 121 (Headquarters of Deutsche Bundesbahn).

is that the potential influence of other government departments, in particular the treasuries, on transport policies is hardly reflected (Henrich-Franke 2009).

## 4. Discussions about International Passenger Services in the Journal of the Association of German Railway Administrations

*4.1 The Railways on the Transport Market – Debates on Integration and Equity*

The railways were well aware of their potential to facilitate travel and thus contribute to spatial integration. In 1871, the new German Reich had been founded, which facilitated movement within its territory, although forms of political integration between its individual states had existed for some time. This was acknowledged in an 1885 contribution to the Zeitung,[8] whose author called upon the railways to support these developments by offering low fares (see below). Similar reports from other countries also reported how movement of people, but also goods, had grown, transforming

> "the whole cultural life of the peoples" and allowing countries "to throw themselves onto the most suitable fields of production and divide work with other nations".[9]

But in contrast to the image outlined above, the railways were not immune to crises and had to learn to handle loss-making operations, in particular during the late 1870s when the German economy was in a difficult situation. Losses were incurred in particular from passenger services. The question whether running them was 'a costly luxury, or at least an investment of doubtful value' was mooted even in the industry's journal.[10] Two strands of arguments emerged from this: to make clear the *railways' contribution to society* at large, and to *integrate the public's concerns* in the industry's strategy in order to become (or at least appear to be) responsive to its needs.

The contribution railways could make to regional development was put forward in their favour. A recurring topic was whether the railways should be extended even further to stimulate regional development and if so, who should organise and plan this. Linked to this is the discussion on the role of the state

---

8    'To maintain family relations will become ever more necessary on the one hand, and ever more difficult on the other as the limits on relationships and interests imposed by the previous small-state structure disappear and give way to a greater freedom to move in all areas of life' ('Zur Reform der Personenbeförderung', Zeitung 1885: 37).

9    'Die Entwicklung und Bedeutung der Eisenbahnen' Zeitung 1899: 826 and 827,

10    'Studien über eine Reform des Personentarifs'. Zeitung 1882: 213, see also ' Der Personentarif d Eisenbahn', Zeitung 1889: 185–7 – ' Die Selbstkosten d Personenverkehrs', Zeitung 1902: 1411.

and the private sector in this industry (see below). Many writers supported an engagement of the state, be it as a planning and coordinating body or as a service provider. In a comprehensive review of railway policy, Heinrich von Wittek (a former Austrian railway minister) praised the efforts of France and Italy who developed comprehensive *national railway plans* already in the late 1870s. He then demanded a 'planned completion of the rail network', because this was

> both a matter of poetic justice, which also allows the less developed parts of the country to participate in the benefits provided by the railways, and a matter of labour and industry policy calling for a balanced occupation of manpower and machinery.[11]

The *role of the state and the private sector* in the rail business is one of the main topics of the debate, more than in the second period where public sector dominance was hardly challenged. Without going into too much detail here – as much has been written already on this subject[12] – it is interesting to note with regard to the international dimension that this issue was discussed in different countries and covered by articles in the Zeitung. The right balance between state and private sector responsibilities was difficult to find, as shown by a review of the railways in Italy published in 1898.[13] At the time, the Italian state had given several concessions to different private companies to build and operate the network. The presentation illustrates well the problems arising both from the difficult terrain, but in particular the insufficient or ineffectively used control mechanisms of this arrangement. The author of the Italian study on which the articles are based puts the blame mainly on a lack of interest by the state to use its influence.

According to the journal articles, there seems to have been a preference for a strong state involvement in the late nineteenth century, as shown by Wittek's article above. But some authors also pointed out that public ownership should not be equal to low fares and continuous subsidies as these would undermine the industry's viability in the long term. The following quotations from 1890 and 1909 sum up this position, which was described as firmly rooted in the German policy tradition:

> In Germany, the plan to shape transport is strongly influenced by the implementation of the idea of state-run railways, this fundamental step which removes the concept of railways as an activity in its own right and elevates these

---

11   'Grundzüge der Eisenbahnpolitik in Anwendung auf die Einzelgebiete des Bahnwesens' (Heinrich Ritter von Wittek), Zeitung 1909: 1541–2.

12   See for example Caron 1988, Fremdling 2003, Gall 1999, Klenner 2002, Ziegler 1996 – with reference to the present e.g. Andersen 1998: 71 et seq., Wink 1995.

13   'Italienische Eisenbahnfragen' (Paul Dehn, referring to a book by the Italian engineer Spera), Zeitung 1898: 1323–5 and 1339–41.

to the height of one of the most important means for planning and implementing
the state's economic objectives.[14]
But for the railways, which generate high operating and capital costs, the
principle of free use, known from primitive means of transport such as streets
and natural waterways, cannot be applied.[15]

Turning to the question who could benefit from the railways' achievements,
*passenger fares* were clearly the topic discussed most. Both the level and structure
of the fare system were controversial issues, where conceptual differences of
opinion existed among the Zeitung's authors. At least some of these can be traced
back to different ideas about society in a wider sense. Regarding the level of
fares, the main issue was whether lower fares would generate sufficient additional
demand to compensate for the lower revenue per passenger. The structure of the
fare system was the second controversial issue. The positive experiences made
with the 'penny postage' (unified fixed tariff for mail) was used as an argument for
the latter solution in several cases,[16] but others rejected it outright because the post
was considered a completely different matter. With regard to the 'internationality'
of the rail industry, it is worth noting that the fare systems differed significantly
across countries, and reference to the experiences made elsewhere were used quite
frequently in the German discussion.[17] The same can be observed in the related
discussion on how much *variety of accommodation* should be provided.

### 4.2 Instruments for International Collaboration and Integration

These examples show that the railway industry was well aware of its importance
for society, but had differing views on how this role should be understood and
fulfilled. Turning towards the role of the railways on the international level, we
can again distinguish two main arguments: the railways' role for international
exchange, and international collaboration within the railway sector itself.

The 'Verein' itself can be seen as one of the first institutions for international
cooperation, and was also referred to as such in several articles. According to a
review of 1886, the organisation had developed into *a truly 'central European
organisation'*, and it was time to convince the Swiss and Italian administrations
to become members as well. The authors also pointed out that other railway
associations should not be considered as substitutes, because

---

14   'Zur Tarifreform', Zeitung 1890: 772.
15   'Grundzüge der Eisenbahnpolitik in Anwendung auf die Einzelgebiete des
Bahnwesens' (Heinrich Ritter von Wittek), Zeitung 1909: 1541–2.
16   'Der Personentarif d Eisenbahn', Zeitung 1889: 185 et seq. – 'Zur Reform der
Personenbeförderung', Zeitung 1885: 37 et seq.
17   For example: 'Der Personentarif d Eisenbahn', Zeitung 1889: 185 et seq. –
'Personenbeförderung auf den Eisenbahnen u d Personentariffrage', Zeitung 1889: 245 et seq.

they actually derive their regulations from the Verein's ones, and so their members have a serious interest in taking part in the development of these original regulations (by becoming member of the Verein, author's note).

Time would come when the Verein, reacting to the growth of its territory, would have to change its name – a sacrifice that should be taken without sorrow.[18] This actually took place in 1932 (Kaessbohrer 1933). The Verein's *standard-setting function* was also mentioned in one of the papers on passenger fares where the author observed that

> once the fare issue has been solved by the German and Austrian railways (the main members of the Verein, author's note), the other Western European administrations will not stick to the current principles ... and the moment will come where one can start to look after international passenger traffic together without intermediaries.[19]

The Verein's role as a coordinating and standardising institution covered several fields, including technical standards, rules for joint use of vehicles and clearing of revenues between its members (for details see Kaessbohrer 1933: 345 et seq.). In passenger transport, the most important activity therefore was the so-called *Vereinsreiseverkehr*, a common fare system for journeys involving its member organisations. These tickets were first issued in 1884, but their spatial availability expanded several times as new administrations joined the scheme. In later stages, non-members of the Verein and non-rail operators (mainly shipping companies in the North Sea and the Mediterranean) also participated. In 1903, a review concluded that

> all countries except Spain and Portugal, from England in the West to Russia in the East, from the Nordkap to the Southern coast of the Mediterranean Sea can be transited by the traveller under favourable conditions.[20]

Without doubt this service can be seen as an important tool to facilitate international travel. As shown by the quotation from Friedrich List in section 2, this had already been part of many railway visionary's thoughts in the earlier nineteenth century. But in the later years, the more conceptual articles in the Zeitung did not devote particular attention to the issue of international connections.

---

18  'Der Verein Deutscher Eisenbahn-Verwaltungen und seine Entwicklung als Mitteleuropäischer Eisenbahnverein', Zeitung 1886: 899.
19  'Die Personenbeförderungsfrage auf den Eisenbahnen und die Personentariffrage', Zeitung 1889: 261.
20  'Zusammenstellbare Fahrscheinhefte', Zeitung 1903: 1049. The evolution of this tariff will be described in more detail in a later publication by the author.

## 4.3 Long Distance and International Service Development

While the debate on the role of the railways and on some topics like fare systems receive continuous attention in the VDEV Zeitung, the number of articles on *service design* (timetables, service standards, vehicle design and the like) is more limited. The most important 'international' issues were without doubt the international express and sleeping car services and their main provider, the Compagnie Internationale des Wagons Lits et des grands Express Européens' (CIWL).

The CIWL had been founded in 1874 by the Belgian George Nagelmackers and was unusual in being a genuine private train operator on the one hand and obliged to collaborate closely with the other rail operators on the other. It ran sleeping car services either as individual coaches or as complete luxury trains, but traction was provided by the rail companies whose territory was served. Furthermore, it operated restaurant cars and later set up hotels, travel agencies and restaurants as well. The transfer of the sleeping car concept from America to Europe by Nagelmackers was without doubt a big improvement for long distance travellers (Widhoff 1965). The company's income was generated from a supplement of the first class fare, while fare revenues themselves were passed on to the train operators providing motive power. In 1898, 18 routes were served by complete CIWL trains, single sleeping cars ran on another 57 routes, restaurant cars on 58 and salon cars on four.[21] Night trains were operated also on other routes by other companies,[22] but the CIWL remained unique in the level of service it provided and as a symbol of a truly European service and of peaceful interaction.[23]

In the above list of CIWL's luxury trains, the *Orient Express* was the first and clearly most famous service. It was opened in stages from its first run in 1883 and developed into a group of services. Before Constantinople could be reached directly, passengers had to transfer to horse carriages in Niš (Serbia), but this was already offered as an integrated service. Innovative approaches were also tried in other areas such as the shifting of customs procedures into the train to speed up onward connections.[24] The locomotives were equipped with cow catchers to reduce problems with cattle in the Balkan area. The company sought to speed up the train as far as possible, and the coordination of timetables across the many company and state boundaries was without doubt a major achievement at the time.

In spite of this, one has to remember that the Orient Express of the pre-war time was a luxury service catering only for a very small elite customer base. The high comfort provided (luxury sleeping cars, restaurant, luggage and staff coaches) led to a very unfavourable ratio of dead weight per paying passenger – an article

---

21 'Internationale Schlafwagen- und Expresszuggesellschaft', Zeitung 1898: 141 et seq.

22 Switzerland for example introduced these with the completion of the Gotthardbahn, cf. Schiefelbusch 2010.

23 'Das 25jährige Jubiläum des Orient-Expresszuges', Zeitung 1908: 868–9, cf. also Dienel 2009 and the contribution of Zaidi in this volume.

24 'Der Orient-Expresszug ', Zeitung 1885: 785 et seq.

from 1890 calculated this as 126: 1 despite assuming a much better occupancy rate than for other trains.[25] In the debate on fare systems described above, this kind of service therefore seemed out of place in the light of other demands made by the public, such as lower fares and faster services.[26]

In terms of service development, some other articles worth mentioning deal with *conceptual issues of service design* and compare the approaches taken in different countries. In 1888, a detailed examination of line speeds operated in England and Germany was presented in reaction to a newspaper claim that trains in Germany could run faster.[27] The authors pointed out that the two railway networks had quite different characteristics, and that if these factors were neutralised, the difference would be far less. Furthermore, faster services in Germany would not necessarily be beneficial for international travellers as it would be difficult to adjust connections to other countries – thus the need for good long-distance connections was recognised despite of the (as share of journeys made) limited number (cf. Schiefelbusch 2010). Another article examined what could be learned from the fact that several railways in England had abandoned first class compartments, while Germany continued to offer three or even four alternative fare levels.[28]

## 5. The Future of International Trains in the European Transport Policy Debate in the 1970s

### 5.1 International Collaboration

Although the railways lost much of their market share and thus also political importance over the second period, optimistic or even *enthusiastic voices* about the opportunities offered by the European integration can be heard throughout this time. They are most evident in the mid 1950s and when some success was observed in the TEE network (see below) and the wagon pool EUROP, and the early 1970s, when several institutions, in particular UIC and the then EEC[29] started to look for possibilities to revitalise the rail sector after a period of decline. Statements of 'borders being obsolete', the need to 'think international' (Ratter 1970: 73) and go beyond national thinking (Sudreau 1970: 113, de Bruin 1972, de Bruin 1971) can be found in particular in the journal editorials. Commentators pointed to progress that had recently been made in European economic integration (Common

---

25    'Ein Beitrag zur Reform der Personentarife', Zeitung 1885: 253 et seq. The ratio of a normal express train (first and second class) was 56: 1m of a passenger train (first, second and third class) 30: 1. The ration for freight trains was given as 1.5: 1 in another paper ('Studien über eine Reform des Personentarifs', Zeitung 1882: 213).

26    'Ein Beitrag zur Reform der Personentarife', Zeitung 1885: 253 et seq.

27    'Unsere Schnellzüge ', Zeitung 1888: 881 et seq.

28    'Verminderung der Klassenzahl im Personenverkehr', Zeitung 1902: 535–7.

29    European Economic Community.

Market). The rail industry needed to adapt, take up this opportunity and learn from it instead of locking itself into a purely national approach of problem solving (Sudreau 1970: 115). To revitalise the railways on a European instead of a national scale would offer significant economies of scale (Lacarrière 1970: 367). As we shall see below, this enthusiasm about the general thrust of European integration was neither universal nor free from concerns and conditions. But it nevertheless went beyond being a mere slogan put forward in high-level debates and similar occasions. A range of measures was discussed to put integration into practice, although there are differences in the nature of the proposals made by more political institutions such as the ECMT[30] and UN ECE[31] and the industry-based bodies.

The UN ECE's document from 1969 discussed two main topics: levels of comfort for passenger rail services and the division of transport tasks between road and rail. Both were covered in good detail. In contrast to what might be expected from a political body, it was detailed regarding service development, but did not consider ways to put these ideas into practice.[32] The ECMT's report from the same year was more comprehensive in the issues considered and also included reflections on the reasons for the railways' current situation.[33] The document was prepared in the execution of the ECMT's work programme and based on consultations with the national delegations. Two main areas of activity were identified: (a) to remove administrative barriers hindering international rail traffic, (b) to get the administrations (departments) to develop incentives for the railways to intensify their collaboration in technical and also commercial matters. Two chapters each were devoted to these two areas of activity, focusing on practical and short to medium-term issues. The report explicitly acknowledged the specific challenges for collaboration in the rail sector by stating that

> the present system of operation is still characterised by each network's focus
> on activities on national scale and that this system never reached the industrial
> dimensions which would be adequate for a truly trans-national scale. [This can
> be] explained without doubt by the singular phenomenon that the railways are
> the only type of transport which does not 'export' its services across the national
> borders. (pp. 25–6)

On the political level, there was far-reaching agreement that this was to change. The most radical proposals went as far as to claim that in a unified Europe, time had come for railways to unite, too, hence a call to merge the national railway companies into some kind of supranational entity (Lacarrière 1970: 369, Sudreau 1970: 115,

---

30   European Conference of Ministers of Transport.
31   United Nations Economic Commission for Europe.
32   All documents in BA KO B108/28977.
33   'Bericht zum Thema der Förderung des internationalen Eisenbahnverkehrs', CM(70)25, Paris 16 November 1970 – BA KO B108/55626. All quotations in this paragraph from this document.

de Bruin 1971: 485, Leber 1971, Leach/McIsaac 1972, cf. Figure 6.1). A less radical strand proposed the formation of new jointly-owned companies to take over the most important (and prestigious) tasks on a supranational level. In the discussion of the EEC's initiative and report on collaboration in the rail industry, the French delegation proposed for example a joint commercial organisation to recapture traffic for which 'the railways, according to their potential, should be the natural carrier'.[34] The proposals of the Dutch delegation went into the same direction and developed some more details on how such a joint organisation serving the main traffic flows could be set up.[35] International container services and fast international passenger trains were suggested as areas of activity. Another, again less novel approach suggested more intense collaboration of the national rail operators on specific tasks in cross-national coordinative bodies, which could either be existing or new foundations. The UIC with its system of sub-institutions, working groups, thematic circles etc. was also proposed as a good forum to develop a more intense collaboration (see below).

In total, the *need for cross-national collaboration* was not challenged in public. But the unpublished sources examined for this analysis show that this enthusiasm

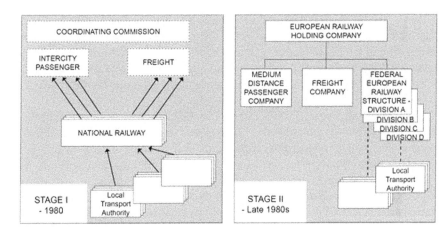

**Figure 6.1    Concept for coordination and merging of Europe's railways**

*Source:* 'Blueprint for a European railway' by Leach, R. and McIsaac, G., Railway Gazette International, April 1972, pp. 131–2.

---

34    Erklärung der französischen Delegation zur Zusammenarbeit zwischen den Eisenbahnen der Mitgliedsstaaten at Council meeting 4 June 1970, BA KO B108/52907, p. 2.

35    Note der niederländischen Delegation hinsichtlich kurzfristig zu ergreifender Maßnahmen zur beschleunigten Verwirklichung einer gemeinsamen Verkehrspolitik. Council meeting 4 June 1970, BA KO B108/52907, see also de Bruin 1971, 1972.

was met with scepticism by important stakeholders. Loss of influence for interests organised on the national level was feared, expressed as warnings against a premature handing over of competencies or duplication of existing structures and by suggesting that *existing cooperative structures* (mainly those provided by the UIC) should be used better before establishing new ones.

In their response to the consultation on the EEC's policy document on the collaboration of Europe's railways, they pointed out that time for a 'European rail company' had not come yet.[36] They conceded the need for better collaboration, but their statement made it clear that they preferred to work together in the existing arrangements as separate entities. A detailed description of current activities, studies and objectives for the coming years was attached to support their claim. Reference was made repeatedly to the UIC's activities, which also would provide a more suitable geographical area of reference than the six EEC states.[37]

This 'pragmatic' approach can be traced back mainly to German interventions, both from the Department of Transport and the Bundesbahn headquarters.[38] The latter's president Oeftering recommended his colleagues in a letter from 15 September 1970[39] to welcome the proposals in principle, but to work against the establishment of new international organisations to promote collaboration. Nevertheless, new working groups and a closer contact of the EEC members' railways within the UIC should be developed. The German Department of Transport was prepared to agree with the French proposals, but quite explicit in its rejection of the Dutch ideas which they considered also an illegitimate attempt to increase the Netherlands' influence on freight traffic to and from the North Sea ports. Any 'brainwashing' of the railways and any attack on their commercial independence would be unacceptable.[40]

The practical scope of this debate on EEC level was to prevent the Community from formulating too detailed and too demanding expectations towards the railways regarding their future international collaboration. The Council discussion reflects the different opinions on how much should be expected from the railways, but also the resolution itself contained alternative wordings on who should be obliged to report on progress. Germany, France and Italy considered it sufficient to ask the railways for an annual joint report. the

---

36   Statement by the grouping of the railways of the 6 EEC Member States to the EEC Council, 8 October 1971, p. 3 – BA KO B108/52907.

37   As before, p. 8.

38   The records in the German archive do not contain detailed information from the other EEC's railways, but minutes from full group meetings show that Italy, Luxemburg and Belgium agreed with the German position in the subsequent debate. Cf. BA KO B108/52907.

39   Cf. BA KO B108/52907.

40   The term 'Gehirnwäsche' (brainwashing) is used in the internal minutes of BMV, unit E2, meeting of transport issues group. Brussels 16 Sep 1970, BA KO B108/52907. On Dutch interests regarding port access see letter of BMV unit E2 to unit A3, 13 July 1970, p. 9 et seq. and undated memo of BMV units A3 and E2 – all BA KO B108/52907.

other countries wanted to oblige Commission and Member States to report on their conclusions, and the Commission wanted the authority to prepare this report itself, make its conclusions and advise the Council.[41] Even this reporting issue was seen sceptically on the German side because it feared that conclusions which might endanger the railways' independence could result from too comprehensive reports.[42] As evident from the second draft of the resolution, the railways' pressure on their counterparts was successful: the final document of 1972 contained numerous requests for the railways to report on or investigate better collaboration in various fields, but placed no formal obligations on them.[43]

To protect their 'independence' from government interventions and control was an interest shared by many European railways, but the German Bundesbahn's insistence on this issue is bewildering for two reasons: first, DB at the same time complained about insufficient financial support from the government for handling its post-war reconstruction and modernisation. Second, the Bundesbahn was actually among those with the least freedom to move and the largest dependence on government approval for its actions (Mäger 1965, in detail Kopper 2007).

### 5.2 Market Environment

The changes in the competitive environment of the railways have already been described briefly in section 2. They also provided an important topic in the industry's reflection on its future potential role. Commentators in the period around 1970 were aware that the railways had experienced a period of decline not only in its market share, but also regarding its perception among users and policymakers. Public interest and confidence could not be taken for granted anymore (Lacarrière 1970).

The *loss of their monopoly position* was acknowledged in the industry, leading to a debate on what the future role of rail could be in a world where road transport continuously gained market share. Comparisons of travel time and fares were for example made in the UIC's corridor studies from 1969–70, where air fares were found to be 1.14–1.51 times those of rail. Competition from air was considered 'very serious'.[44] Another strand of arguments which in the end worked against facing international collaboration in a pro-active way was the reliance of railway managers *on regulation and* other types of *state intervention* as a means to combat losses of market share (e.g. UIC 1951, UIC 1953, cf. Kopper 2007).

---

41    Entwurf für eine Entschließung des Rates über die Zusammenarbeit zwischen den Eisenbahnen (R/2437/d/70) and minutes of Council discussion on 8 October 1970, BA KO B108/52907.

42    As above, p. 4 of minutes.

43    Draft translation in internal document, BMV unit E2, BA KO B108/52907.

44    Synthesis report of GPC 'Future of international passenger transport', January 1970, p. 4 – DOC UIC 1375.

The competitive situation clearly influenced the debate on '*promising markets for rail*'. Sometimes this was expressed positively as 'areas of competitive advantage' (Lacarrière 1970: 367), but occasionally also by an explicit acknowledgement that certain activities should be given up.[45] Such considerations were developed at about the same time both on European level and within the different countries and often led to new policy guidelines or reports on future rail and transport (e.g. the well-known Beeching report in Britain, the Nora report on the French railways and the transport policy programme developed in Germany after the new Minister Georg Leber took office in 1969 (cf. Kopper 2007: 30, Zeilinger 2003: 98 et seq.). Long-distance traffic, container trains and some other parts of freight transport, and local passenger services in agglomerations were usually recognised as markets with potential for rail. Even within these segments, however, a need for some consolidation and withdrawal of unprofitable parts was considered necessary.[46] While the discussed reductions affected mainly short-distance and rural traffic, this does not mean that the long-distance sector – was generally seen as an interesting market. International services were not necessarily considered explicitly.

*5.3 Service Development*

Despite of the negative developments in rail's market share, the period under review was also a time of *progress*. Steam traction was substituted by electricity and combustion engines, and at least the mainline infrastructure modernised. These technical issues were also discussed on a supra-national level, even though common standards could not always be achieved.

One 'visionary' strategy to offer something new on this market clearly is the *Trans Europ Express* or TEE network of fast premium train connections between different European states. This was introduced with the summer timetable of 1957, following an idea first set out by the president of the Dutch Railways Den Hollander in 1953 and echoing similar 'European' network building activities in the road sector (Schipper 2008). His idea was to connect the most important cities in Europe with a network of fast, direct passenger trains of a premium standard, to create (after the formation of the freight wagon pool EUROP, the financing body Eurofima and others in the preceding years) yet another symbol of collaboration among Europe's railways (anon. 1957, de Bruin 1967, Ratter 1970). The seven administrations that agreed on the TEE's introduction[47] formulated a set of quality standards for this service: trains should be electrical or diesel multiple units,

---

45   UN ECE, Inland Transport Committee, sub-committee 'Railways', Secretariat's note – W/TRANS/SC2/220, p. 3 et seq. – BA KO B108/28977.

46   UN ECE, Inland Transport Committee, sub-committee 'Railways', Secretariat's note – W/TRANS/SC2/220, p. 3 et seq. – BA KO B108/28977.

47   Belgium, France, Federal Republic of Germany, italy, Luxemburg, Netherlands, Switzerland.

they should have a capacity of at least 100, a restaurant service and high-quality interior in terms of spaciousness, insulation and air conditioning (Huber 1957, Stöckl 1971, de Steenwinkel 1957, de Bruin 1967). TEE trains conveyed first class only and although most administrations built their own rolling stock, they all agreed on a common colour scheme and logo. The TEE network was to expand further in the following years, also into countries not yet served, but gradually shrank from the early 1970s onwards until services officially ended in 1987 (for details see Mertens 1987, Stöckl 1971, cf. Figure 6.2)

In terms of integration and standardisation, it is worth noting that the original concept underwent significant changes both before it was first implemented and during its existence, making these 'standards' very flexible. The most important deviation from the original concept was the provision of services by the national operators on their own (with bilateral technical collaboration), whereas Den Hollander originally had envisaged TEE trains to be run by a new 'European' company owned jointly by the national administrations. In the debate that followed both on the political level and the companies, it quickly became clear that the proposed new type of service was welcomed, but the formation of a separate company was considered too difficult and not necessary (Kopper 2010). Den Hollander argued that only the separate company would pursue the associated tasks of standardisation and integration with the effort required, but in the end remained isolated in this view.[48]

The topic was picked up again in the late 1960s in the debate on the railways' future on the passenger market. In these discussions on a *new 'standard' model for international passenger services*, the TEE concept was still an important, apparently unchallenged model,[49] although first route closures were to be discussed almost at the same time. According to the documents, working group members agreed that a similar model, but with a wider commercial outreach, in particular regarding non-premium customers, would be desirable. But there was no agreement on how this model could look like. At the time of these discussions, the 'Intercity' (IC) brand had been introduced for domestic fast, high-quality trains in Germany (Gall 1999: 360 et seq.), Britain (Shin 2010) and some other countries. The first German IC network (introduced May 1971) consisted of several fixed routes where first class-only trains operated in two-hourly intervals with consistent speeds and stopping patterns. It was deemed a success, but its true potential only became apparent when the service was reorganised in 1979 to offer also second class and more frequent (hourly) services.

In the UIC's discussions of the early 1970s, this concept was one of the options discussed, but participants could neither agree on it nor did a convincing alternative

---

48   Minutes of different meetings of railway companies summer – autumn 1954, in particular Utrecht 14 September 1954, p. 4, all BA KO B121/473.

49   Note of the GPC to the Pushing Group on issues raised by Mr Wichser (SBB), attachment 4 to agenda of Pushing Group meeting 1 June 1970 (original German), DOC UIC 1375.

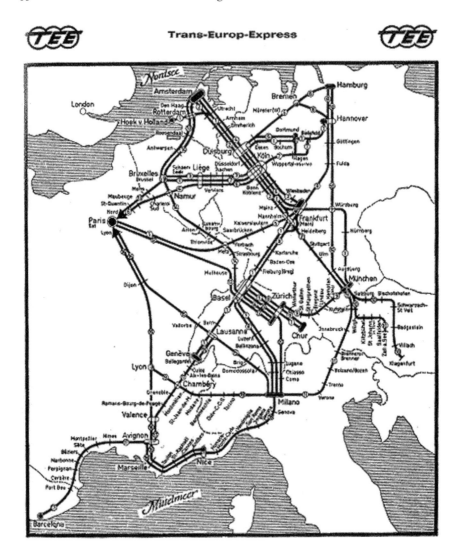

**Figure 6.2    The TEE network in the late 1960s (Stöckl 1971: 5)**

emerge. The DB representatives sought to avoid changes to their concept, in which much had already been invested at the time. Furthermore, other railways had started to use the term 'Intercity' as well, but with different connotations, which seemed justified based on their national requirements, but were impossible to harmonise.[50] The Swiss representative preferred to speak of 'high-quality trains'

---

50    Minutes of 4th meeting of GPC, Paris 29 April 1970, pp. 1–9, DOC UIC 1375.

category A and B' as new service types underneath the TEE, thus avoiding the word Intercity. Yet another approach was taken by the French SNCF who had started to denominate its premium domestic trains as TEEs – against the core concept developed in the 1950s. Although the TEE had become a rather vague concept at that time,[51] Louis Armand (former SNCF and UIC president and an influential figure in Europe's railways) and the managing director of Swiss SBB Wichser expressed scepticism towards any idea of establishing a second brand for international passenger trains in parallel to the TEE.[52] In the end, no agreement was reached on this point.

The group's work was criticised also by insiders. Some initiatives to analyse the decline of international traffic in the 1960s had led to insufficient responses by the member administrations, and efforts to obtain precise data were not taken far enough.[53] The wide range of topics combined poorly with the restricted time and resources its members could devote to their work.[54] On the other hand, the coordination between the different UIC bodies was criticised as leading to unnecessary duplications. In February 1971, the Pushing Group's president Weber from the Belgian SNCB resigned because of the lack of support and progress.[55]

Already in 1967 had a *'Commission for Prospective Research'* been set up following a decision by the UIC for conceptual long-term planning (anon. 1970b). It submitted its report to the Executive Committee in March 1970 (UIC Recherche Prospective 1970, anon. 1970a). Under the presidency of the French SNCF, a survey of member administrations' activities and ideas was conducted. Regarding the status quo analysis, a convergence of ideas and tendencies regarding the improvement and development of the rail offer was noted. However, compared to the fast development of international transport by other modes, the results achieved in the rail sector seem less impressive. It seemed that the European railways risked to remain a simple juxtaposition of national systems. The railways' 'prise de conscience' of towards international business was insufficient.

---

51   More details on the various services and modifications in Mertens 1987.

52   Sources for this section: Minutes of GPC meeting 28 April 1971, Minutes of PG meeting 7 June 1971 (originals German), DOC UIC 1372 .

53   Minutes of Commercial Conference on Passenger Services, section on the analysis of international passenger traffic (original French), Basel, 23 September 1968, p. 146, DOC UIC 3902.

54   Method of working of GPC, attachment B to agenda of 14 December 1970 (original German), DOC UIC 1377.

55   Letter of Mr Weber to Mr de Bruin as acting president of the Pushing Group, attachment C to PG meeting 7 June 1971 (original German), DOC UIC 1372.

## 6. Analysis and Outlook

If we compare these findings, the role of the 'international dimension' in the two periods shows interesting differences. Simplifying as far as possible we may conclude: In the first period, *no common vision* – in the form of a commonly discussed and agreed document – emerged from the documents studied. No call for such a vision could be identified either. But on the other hand, the '*internationality of railways*' was practised in various ways, in particular by the ongoing expansion of network and services and the necessary evolution of cooperation within the industry. This arguably kept pace with the evolving needs of the public. The fact that rail development was to a larger extent driven by profit-oriented both public and private interests certainly contributed to this. This may explain why international issues were less controversial. The references made to the situation in other countries in the Zeitung's articles show that, as a frame of reference and source of ideas, international exchange was lively and useful.

In the second period, not one but several visions were put forward (from technical collaboration via service standards to a pan-European rail operator), but on the whole the outcome was less impressive. In the 1950s, railway industry leaders were caught in the enthusiasm for European integration as every other industry and developed the TEE brand as their contribution. But in parallel, rail lost ground on the transport market and also political attention. By the late 1960s, the situation had become so unsatisfactory that political leaders returned to the argument and demanded change, as the quotations in section 5 show. They had some of the railway managers as their allies, and in the early 1970s the first plans for a *European* network were presented at last. However, the references in this paper also show that the rail industry did not agree on the way forward in detail. This would not have been the main problem, but it is also evident from the sources that the future vision for the industry was closely linked to influence and control. Influential parts of the industry tried to prevent the formation of new, competing bodies and to keep the role of outsiders to a minimum. Thus there may have been *visions, but no visionaries* to implement them.

Regarding the *instruments and strategies used to achieve integration*, cooperative approaches clearly dominated at least since the 1870s (Henrich-Franke 2007). The many coordination forums like the European Timetable Conference and the strong involvement of the VDEV and especially later the UIC in the practical work of discussing the state of the art, doing research and developing standards and harmonised policies is a key feature of the rail industry. It is arguably a reflection of both the need to achieve a precise common understanding on many technical matters (Schiefelbusch 2010) and the firm links of most rail operators to a network with clear geographical boundaries and hence limited overlap and competition. In this situation, market-driven standardisation cannot assume a major role as a 'market' on which different providers really compete does not exist.

Political bodies also do their part, in particular for financing infrastructure projects, covering deficits and as legislative or regulatory powers. In the first period, opinions on the level of public sector involvement were still divided, but calls for a greater role at least in setting a framework were already made. In the second period, the national governments had – both in this role and as owners of the rail companies – the powers to do this. But unfortunately they often did not fulfil this role, either because other modes and issues were more important or because they had handed the task of strategic thinking back to the rail companies.

For the topic of this chapter, the above shows that it would be too much of a simplification to associate the first period only with private enterprise, market-based integration and harmonisation and the second with public-sector dominance, as is sometimes done. There was much awareness of rail's importance for society and attempts to secure public influence even in the first period, as section 4.1 has shown. These debates may be called a 'competition of models', but such competition can also be found in period 2. Here, the resistance the railway administrations showed against the political calls for improved collaboration shows that the industry was far from a purely hierarchical system with the political leaders at the top. The argument of the administrations' independence as 'enterprises' was also used to counter attempts for political control (cf. section 5.2). Nevertheless, the state is clearly more evident as a stakeholder in rail development in the second period.

What do these thoughts *tell us for today*? Activities on European level for the integration of transport infrastructure and services have without doubt increased significantly in intensity since the 1970s. There is no space here to discuss details of these developments, but two items which connect to the issues described above shall be mentioned: First, in spite of the common interest in high-speed rail across Europe, the concrete projects and technologies were first developed on the national level. Although there was cross-national exchange of information, the solutions (in particular the new train units) became known as *national* achievements and symbols. High speed rail may be a common idea, but not a common technology (Bouley 1981). To integrate these into an international vision of European rail transport was therefore not a priority. It took another 20 years until first steps in this direction were taken in the late 1990s with the first proposals for Trans-European Networks, a process brought forward on the one hand by a further increase in cross-border traffic and by the revived interest of the EU in a 'common' transport policy. Even with these unifying forces, the expectations towards a European high speed rail policy still differ across the continent (Andersen/Eliassen 1998: 71 et seq.).

Second, the idea of a '*single European railway*' has been picked up again in the EU's transport policy over the last years. However, this is today more associated with removing technical and administrative barriers and liberalisation of the industry than with the creation of one unified company. In this context, f barriers is a prerequisite for liberalisation and – hopefully – the f the railways to work. As described above, such harmonisation  been pursued in a framework of collaborative decision making.

The participating administrations discussed together, but otherwise worked side by side in geographically confined areas.

It remains to be seen if the industry's system and culture of collaboration can be maintained and improved when it is expected to become more market-driven and competitive at the same time. While competition may be a good way to bring in new ideas, there is a risk of losing what has been achieved as well. EU policy has in part acknowledged this by setting up new institutions like the European Railway Agency (ERA) or the national rail regulators who take over some of the tasks previously done by the rail companies themselves. According to its first evaluation, the ERA has so far been dealing mainly with technical aspects of harmonisation. The development of a 'a genuine European railway culture' (Steer Davies Gleave 2011: 58) has not been an explicit task so far, but according to the report, it could, and should, become more active in this field, for instance through shared understanding and wider use of common methods of working' (Steer Davies Gleave 2011: 59–60). The report makes some further suggestions, but does not discuss the links between these new EU-driven initiatives and the industry's established institutional and administrative system of collaboration.

This sequence repeats railway evolution in the nineteenth century, where the essential technical issues were also standardised first. Based on this, a common understanding of working together seems to have developed towards the end of the century, in spite of quite marked differences of opinion, as the analysis above has shown. From this perspective, it seems interesting to take a closer look at how this collaborative spirit could develop and examine what can be transferred to today.

**References**

Andersen, S.S. (ed.) 1998. *Making policy in Europe. The Europeification of National Policy-making.* Reprinted. London: Sage.

Anon. 1957. Une page nouvelle dans l'histoire du Chemin de fer: le Trans-Europ-Express. *Bulletin de l'Union Internationale des Chemins de Fer*, vol. 28 (6), 177–8.

Anon. 1970a. Le Comité de Gérance de l'Union Internationale des Chemins de fer. *Rail International* (4), 273–4.

Anon. 1970b. Rapport de gestion du Comité de Gérance à l'Assemblée Générale de l'IUC. *Rail International*, 186–8.

Bouley, J. 1981. TGV und UIC. *Schienen der Welt*, 581–6.

Bruin, M.G. de. 1967. Zehn Jahre TEE. Die Zukunft des europäischen Qualitätsverkehrs. *Die Bundesbahn*, 303.

Bruin, M.G. de. 1971. Zu einer engeren Zusammenarbeit zwischen den europäischen Eisenbahnunternehmungen. *Schienen der Welt*, 479–91.

Bruin, M.G. de. 1972. Erfahrungen von 50 Jahren Zusammenarbeit als Orientierung für die Zukunft. *Schienen der Welt*, 733–40.

Burri, M. (ed.) 2003. *Die Internationalität der Eisenbahn, 1850–1970.* Zürich: Chronos (Interferenzen, Bd. 7).

Caron, F. 1988. The evolution of the technical system of railways in France from 1832 to 1937, in *The Development of Large Technical Systems,* edited by R.H. Mayntz. Frankfurt Main: Campus, 69–103.

Dienel, H.-L. 2009. Die Eisenbahn und der europäische Möglichkeitsraum, 1870–1914, in *Neue Wege in ein neues Europa. Geschichte und Verkehr im 20. Jahrhundert,* edited by R. Roth. Frankfurt am Main: Campus, 105–23.

Fremdling, R. 2003. 'European Railways 1825–2001, an Overview', in *Jahrbuch für Wirtschaftsgeschichte, Neue Ergebnisse zum NS-Aufschwung,* edited by J. Ehmer; R. Fremdling; H. Kaelble. Köln: Akademie Verlag, 2003/1, 209–21.

Gall, L.P. (ed.) 1999. *Die Eisenbahn in Deutschland. Von den Anfängen bis zur Gegenwart.* München: Beck.

Henrich-Franke, C. (ed.) 2007. *Internationalismus und Europäische Integration im Vergleich. Fallstudien zu Währungen, Landwirtschaft, Verkehrs- und Nachrichtenwesen.* Baden-Baden: Nomos.

Henrich-Franke, C. 2009. Die Eurofima: Standardisierungsmotor zwischen ökonomischer Notwendigkeit, europäischer Wünschbarkeit und nationaler Realisierbarkeit, in *Standardisierung und Integration europäischer Verkehrsinfrastruktur in historischer Perspektive,* edited by G. Ambrosius et al. Baden-Baden: Nomos.

Huber, E.W. 1957. Zur kommerziellen Gestaltung des Trans-Europ-Express-Dienstes. *Die Bundesbahn,* 598–601.

Kaessbohrer, A. 1933. Der Verein Mitteleuropäischer Eisenbahnverwaltungen. *Archiv für Eisenbahnwesen* (56), 12–34, 345–80.

Klenner, M. 2002. *Eisenbahn und Politik 1758–1914. Vom Verhältnis der europäischen Staaten zu ihren Eisenbahnen.* Wien: WUV (Dissertationen der Universität Wien; 81).

Kopper, C. 2007. *'Die Bahn im Wirtschaftswunder'.* Deutsche Bundesbahn und Verkehrspolitik in der Nachkriegsgesellschaft. Frankfurt/M: Campus (Beiträge zur historischen Verkehrsforschung, 9).

Kopper, C. 2010. Die internationale Zusammenarbeit der Deutschen Bundesbahn, in *Internationale Politik und Integration europäischer Infrastrukturen in Geschichte und Gegenwart,* edited by G. Ambrosius, C. Neutsch and C. Henrich-Franke. Baden-Baden: Nomos, 213–31.

Lacarrière, P. 1970. Zukunft d Eisenbahn in Europa. *Schienen der Welt,* 369–75.

Leach, R/McIsaac G. 1972: Blueprint for a European railway. *Railway Gazette International,* 129–32

Leber, G. 1971. Die Eisenbahn und ihre Zukunftsprobleme. *Schienen der Welt.*

Mäger, F.-O. 1965. Die Deutsche Bundesbahn als Verkehrsunternehmen. *Gegenwartskunde-Zeitschrift für Wirtschaft und Schule,* vol. 14, 111–22.

Mertens, M. 1987. *Trans Europ Express. TEE.* Düsseldorf: Alba.

Ratter, J. 1970. The continental concept of railways. *Rail International,* 73–4.

Rotteck, C. von. 1837. *Staatslexikon oder Enzyclopädie der Staatswissenschaften.* Bd. 4. Unter Mitarbeit von Carl Welcker. Altona.

Schiefelbusch, M. 2010. *Collaboration and Common Visions in the Development of An International Rail Network in Europe.* Presentation at Tensions of Europe conference, Sofia, 19 June 2010.

Schiefelbusch, M. 2013. *Trains across Borders. Comparative Studies on International Collaboration in Railway Development.* Berlin.

Schipper, F. 2008. *Driving Europe. Building Europe on Roads in the Twentieth Century.* Amsterdam: Amsterdam University Press.

Shin, H. 2010. *InterCity: The Regeneration of Britain's Railways, 1950s–1970s.* Presentation at 8th T2M Annual Conference. New Delhi, 2–5 December 2010.

Steenwinkel, M. de. 1957. L'organisation du service du Trans-Europ-Express. *Bulletin de l'Union Internationale des Chemins de Fer, vol.* 28 (6), 180–86.

Steer Davies Gleave. 2011. *Evaluation of Regulation 881/2004.* London. Available at: http://ec.europa.eu/transport/evaluations/doc/2011_era-evaluation-881–2004. pdf.

Stöckl, F. 1971. *Trans-Europ-Express.* Der Werdegang des TEE-Betriebes. Augsburg: Rösler + Zimmer.

Sudreau, P. 1970. Le rail et l'Europe. *Rail International,* 113–16.

UIC. 1951. *Die Lage der europäischen Eisenbahnen.* Bundesarchiv Koblenz, B121/493.

UIC. 1953. *Die Lage der europäischen Eisenbahnen. Maßnahmen aus der Denkschrift des UIC vom Februar 1951, Anregungen für in Aussicht zu nehmende neue Maßnahmen.* Bundesarchiv Koblenz, B121/493.

UIC Recherche Prospective. 1970. La recherche prospective dans les chemins de fer. *Révue Générale des Chemins de Fer,* 537–53.

Widhoff, A. 1965. Rollende Hotels in 26 Ländern. *Die Bundesbahn,* 777–781.

Wink, R. 1995. *Verkehrsinfrastrukturpolitik in der Marktwirtschaft.* Eine institutionenökonomische Analyse. Berlin: Duncker & Humblot (Schriftenreihe des Rheinisch-Westfälischen Instituts für Wirtschaftsforschung, Essen, 59).

Zeilinger, S. 2003. *Wettfahrt auf der Schiene.* Die Entwicklung von Hochgeschwindigkeitszügen im europäischen Vergleich. Deutsches Museum – Beiträge zur Historischen Verkehrsforschung. Frankfurt am Main: Campus.

Ziegler, D. 1996. *Eisenbahnen und Staat im Zeitalter der Industrialisierung.* Die Eisenbahnpolitik der Deutschen Staaten im Vergleich. Stuttgart: Steiner (Vierteljahrschrift für Sozial- und Wirtschaftsgeschichte. Beihefte, Nr. 127).

# Chapter 7

# Functionalistic Spill-over and Infrastructural Integration: The Telecommunication Sectors

Christian Henrich-Franke

## 1. Introduction

The international regime for the regulation of the telecommunication sectors in Europe has grown in complexity since its establishment in the nineteenth century. Institutional arrangements and fields of activity were permanently expanded. Today, the telecommunication sector is one of the most important backbones for a globalised society.

The aim of this chapter is to explain the growth and to compare pan-European regime changes in the telecommunication sectors in the nineteenth and twentieth century by making use of the functionalistic and neo-functionalistic thinking. Of particular interest are the functionalistic key concepts: the strong focus on technocratic governance and the evolutionary development of interstate relations. The changes of the telecommunication regime will be here considered as results of spill-over effects, which either occurred subsequently or in direct connection with the regime changes. All three modes of telecommunication will be taken into account: the electric telegraph, telephony and radio.

This chapter is based on the idea of a twofold concept of integration. Integration means either the standardisation of different economic parameters or the institutionalisation of forms of governance. When telecommunication infrastructures are connected and utilised internationally, a multitude of issues need to be coordinated and regulated. These can be classified into certain areas of activity, the majority of which is associated with the standardisation of different parameters:

- *technical standardisations* like telegraph cables or radio equipment;
- *operational standardisations* like codes, signals or operation times;
- *administrative standardisations* like clearings or organisational processes; and
- *standardisations of tariffs*.

In order to differentiate degrees to which telecommunication networks are integrated, this chapter will make use of three levels of standardisation:

- *high-level standardisation*, where the networks and equipments are interoperable;
- *medium-level standardisation*, where networks are interconnected (adjusted), but where terminal equipment is incompatible; and
- *low-level standardisation*, where gateway technologies interconnect technically distinct networks at the borderlines.

Another indicators for a high-level (or medium-level) integration are the standardisation of telecommunication policies or common enterprises, for example, if several (national) service providers pool their resources and establish an international service like a telecommunication satellite.

Historical research has dealt quite a lot with the international integration of infrastructures in the last years. Many studies on individual infrastructures were published, however, the telecommunication sectors (telegraph, telephone and radio) have not been a prominent part of it. Exceptions are the dissertation by Leonard Laborie on 'La France, l'Europe et l'ordre international des communications (1865–1959)' (Laborie 2008), a dissertation by the author himself on the global regulation of radio after 1945 (Henrich-Franke 2005) and a limited number of articles (Ahr/Benz/Tölle 2010) dealing with isolated aspects of international telecommunication. Recently, the journal *Historical Social Research* has published a special issue on global telecommunication in the late nineteenth century which had a strong focus on information flow and the creation of spaces by telecommunication (Wenzlhümer 2010). From a methodological point of view articles written by Johan Schot are interesting for our topic, because he applies neo-functionalists´ thinking to the history infrastructure integration in the European Union (Schot 2010).

In order to reach its aim, this chapter will start with some reflections on functionalistic and neo-functionalistic thinking with a particular focus on the concept of spill-over. In a second step, the development of the telecommunication regime since the nineteenth century is described. This chapter is subdivided into several parts according to the different regime changes that will be compared and explained. Each part will first consider developments preceding the regime change and then focus on the change itself. Finally, a conclusion will be drawn.

## 2. Functionalism and the Concept of Spill-over

Functionalism or neo-functionalism as its enhancement is not a real theory of international relations. It is rather an approach that aims at explaining the underlying logic of the development of international cooperation. It starts from the assumption that in the nineteenth and twentieth century the nation state is no longer capable of fulfilling human needs. These can to an increasing number only be satisfied in a transnational context. Transnational institutions are seen as the more efficient providers of welfare compared to national governments. Functionalists give preference to technocratic forms of governance within such transnational institutions. The 'government of men' should be replaced by the 'administration of things' because politicians are seen as people maximising power, whereas technocrats have a stronger focus on human welfare (Rosamund 2000, 34).

The intellectual father of functionalistic thinking was the British political scientist David Mitrany (Mitrany 1943). In the 1930s he already argued that the dissolution of governance for domestic purposes has a strong impact on the international order. The growing 'material interdependencies' of states would force the protagonist to rearrange the institutional arrangements. They would – in an evolutionary process – find transnational solutions to solve the problems of public management, distribution, welfare and communication. For Mitrany, this process would begin to emerge informally with the interstate contacts in the field of technology. Shifting patterns of (human) needs would put pressure on the protagonist to bring international institutions in line. Subsequently, ever more institutional solutions would be established. Mitrany expected that functional organisations, which are founded as a consequence of rational preoccupation with human needs, would produce remarkable benefits. Such benefits would create the conditions for the expansion and reproduction of functionalistic organisations. New institutions are then established as a consequence of an increasing pressure on the international arrangements that results from shifting patterns of needs (Giering 1997). Functionalists put this mechanism in a short formula: form follows function.

Mitrany's idea has often been applied to the European Union. Since Ernst B. Haas guiding book *The Uniting of Europe* (Haas 1958), which was published in 1958, the neo-functionalist approach has been used to explain the building of the European Communities. Neo-functionalism underwent many ups and downs. After having been declared dead by Haas himself in the 1970s (Haas 1975), when European integration made no progress, neo-functionalism was revived in the 1990s by authors like Alec Stone Sweet, Wayne Sandholtz and others (Sandholtz/Stone Sweet 1998). In this chapter, however, the European Union's evolution is just of secondary interest.

Neo-functionalists theorise three mechanisms that drive forward the functionalistic integration process: positive spill-over, technocratic automaticity and the transfer of domestic allegiances (Rosamund 2000, 50–54). The last one is directly connected to supranational governance within the European Union and is therefore not of interest in our context. The concept of 'positive spill-over' is however a more general mechanism. It consists of several elements which will be used in this chapter in order to explain regime changes in the telecommunication sectors. Here, we can distinguish between an 'economic spill-over' and an 'institutional spill-over':

- *Economic spill-over*: In general, economic spill-over means that the integration of particular economic sectors will create functional pressures on the integration of related economic sectors. In the long run, this results in a gradual and progressive entangling of national economies. Applied to telecommunication infrastructures and our concept of integration by standardisation we can distinguish between three different kinds of spill-over:

- A spill-over from one type of standard to another type of standard.
- A spill-over from one type of standard to the same type of standard across different telecommunication infrastructures.
- A spill-over from standards to a common undertaking or to policy aspects within the same or across different telecommunication infrastructures.

• *Institutional spill-over*: Deepening economic integration requires the creation, application and interpretation of rules. Therefore institutions are established. Ben Rosamund put it in a simple formula: 'Political integration is an inevitable side effect of economic integration' (Rosamund 2000, 36). Deeper economic integration is in need of regulatory capacity. Applied to infrastructures and their standardisation, we can distinguish between two different spill-over:

- Ongoing standardisation and/or economic spill-over put pressure on the protagonists to establish international institutions to negotiate standards.
- Institutional arrangements are transferred (or copied) across different telecommunication infrastructures.

Next to the functionalistic mechanism of spill-over, the concept of 'technocratic automaticity' represents a helpful tool as well. The idea is that the participants in international institutions and the actors of international institutions take the lead in sponsoring further integration in order to become more powerful and autonomous.

## 3. Changes in the Telecommunication Regime

### 3.1 The Telecommunication Regime's Pre-history

The electric telegraph had been limited to the regional or national space in its beginning in the 1830s. Often replacing some more rudimentary systems of optical telegraphy, the electric telegraph was restricted in its use to state – and in particular to military – purposes. At the end of the 1840s, network operators, who were mostly state administrations at that time, decided to open telegraph lines for private (economic) users. They aimed at making telegraph lines more cost-effective. This decision subsequently increased the demand for trans-border communication. Industrialisation and the international economic relations pressed ahead network expansion and made international standardisation a topic. The major challenge at that time was the technical diversity between different states and regions. Each telegraph had taken its own technical development path. In addition, operational parameters like coding systems required harmonisation and upon the tariffs for trans-border communication had to be agreed (Wobring 2007, 85–92).

The need for standardisation soon resulted in a number of bilateral agreements that tried to standardise operational and technical parameter in order to make trans-border communication possible. Even tariffs were tackled in the negotiations, but they turned out to be a difficult matter. The agreements succeeded in making the networks inter-connectable. However, they were not technically connected. Instead, the telegraph stations at the borderline received a message from one network, wrote the message down and then send it further to the neighbouring telegraph network. Technically separated networks were interconnected by an operational standard. This procedure, however, was quite ineffective and could not satisfy the growing demand for communication. Therefore, in 1851 the first gateway technologies, which were based on the Morse telegraph, were introduced. That led to even technically connected telegraph networks. Network operators were enabled to send messages directly from one network into the other.

We can observe a first 'institutional spill-over' of limited geographical scope when the Deutsch-Österreichischer Telegraphen Verein (DÖTV) was established in 1850 as the first international telecommunication organisation. It should centralise the negotiations on standards and rules for trans-border telegraphy. The DÖTV agreements were soon used as a model across Europe and adopted within other agreements. In the 1850s, standards for electrical telegraphy often spread across Europe in a non-cooperative way. Even the Western European Telegraph Union, which was established by a number of south-west European countries in 1855, simply adopted the DÖTV agreement. Nevertheless, the standards were just some very basic operational or technical provisions (low-level standardisation). They did not seriously harmonise individual components of the telegraph networks. They were rather at the lowest level to enable trans-border communication.

The economic conditions at that time were very favorable for such a way of standardisation.

(1) Electrical telegraphs were in an early stage of development. The few installations were comparatively easy and inexpensively harmonised as the sunk costs were relatively low. Therefore, it was easy to replace different technologies. When then the networks rapidly expanded in Europe in the second half of the 1850s (new lines and multiple-gated lines), this was done on the basis of low-level standards regarding the operation and the technical equipment.

(2) An ever growing demand for telegraph communication and technical innovations accelerated the pan-European standardisation.

(3) Some countries at the European periphery, which lagged behind in the development, built their networks according to the basic standards.

It needs to be underlined that telecommunication's development was not planned foresightedly in this early period. Instead, pragmatism was guiding the protagonists. They were surprised by the (technical and economic) speed of the development and rather reacted than acted. Hardly anyone anticipated the evolution of telegraphy towards a mass phenomenon. Long-term planning or conceptual considerations were as unknown as a coherent telecommunication policy. This was quite usual in infrastructure sectors even outside the telecommunication sector. In the nineteenth century even the differentiation between interconnectivity and interoperability is mostly an analytical one which did not guide the protagonists' thinking.

## 3.2 The Regime Change in the 1860s and 1870s

Tariffs proved to be the most difficult aspect of standardisation in the early years of the electric telegraph. Neither within the DÖTV nor within other bilateral or multilateral agreements had it been possible to negotiate a tariff system (Reindl 1993). The DÖTV even threatened to break apart because of a missing harmonisation of tariffs in the early 1860s. The network operators, which by majority were monopolistic national Postal, Telegraph and Telephone (PTT) administrations, wanted to decide on their tariffs independently since electric telegraphs mostly were a comparatively unprofitable business. Often, it was the governments that kept an eye on tariffs. Without any European tariff system the rates were simply summed up in trans-border telegrams. The tariffs' diversity resulted in complaints from the industry and subsequently inspired the French government to call an all-European conference to decide on a European tariffs system at Paris in May 1865 (Wobring 2007, 106–9).

The Paris conference went beyond a European tariff system and founded the International Telegraph Union (Union Internationale de Télégraphique – UIT) to which nearly all European governments acceded. They adopted uniform operating instructions and laid down a European international tariff as well as accounting rules. The delegates simply led the regulations of the bi- and multilateral agreements for the electric telegraph merge into one central agreement. Standards which had spread uncooperatively over the political market were now multilateralised. Network effects and the economies of scale and scope were widely accepted. Apart from that, the agreement had no further effect. Governments simply consented to already established regulations at a low-level which left them freedom to independently standardise national networks and equipment.

Also, the European tariff system turned out to be a low-level standard. It was merely a list of the existing tariffs than a unified single tariff system. The enlargement of the palette of standards with tariffs was the actual 'spill-over'. For the first time, the European states succeeded in drawing up a European tariff system. The pressure from the economy was as important for this step as the benefits deriving from the cooperative standardisation of operational, administrative and technical parameter.

The foundation of the UIT was not a real 'spill-over', but rather an administrative act, that had no consequences for the processes of standardisation. Nevertheless, the expansionist development of the electric telegraph had an enormous impact on the character of the standards. On the one hand, operational, administrative and technical aspects still dominated. On the other hand, the quality and quantity of standards changed. The growing complexity of the networks and technical innovations made ever greater demands for the standardisation. International interconnectivity needed permanent adjustments, even though it remained at a low-level of standardisation.

The differentiation of the standards put an enormous pressure on the UIT's institutional arrangements and initiated the regime change. In 1875 at a conference in St. Petersburg, the UIT already went through a decisive revision that reflected the changes in the characteristics of the standards. The UIT agreements were divided into a Convention, which laid down the organisational structure and fundamental rules for telegraph communications, and the Telegraph Regulations, which contained more variable standards that needed to be adjusted to the technical and operational development of the telegraph. The organisation was split up into an intergovernmental plenipotentiary conference which could decide on the Convention and conferences of PTT-administrations to decide on the Regulations and the table of telegraph rates. The Convention solely defined the basic principles of trans-border information flows, like the secrecy of telecommunication, the responsibility for information flows, the protection of installations, the provision of channels and installations to ensure rapid or uninterrupted exchanges. The basic principles were stated more precisely within the Telegraph Regulations. This division initiated an enduring process of turning international telecommunication governance into a technocratic affair. Governmental representatives were only consulted when the Convention was revised. If we put this in functionalist terminology, than we see all signs of a replacement of the government of men by the administration of things. Nevertheless, standards were solely coordinated at periodic conferences at the international level. Technical and operational developments were separately studied within the individual states. A multi-level system of standardisation had emerged (Convention, Regulation, bilateral agreements) that was characterised by a top-down increase in the level of standardisation.

The decisions made in 1875 fixed a pattern of integration that had emerged out of the numerous agreements in the 1850s and 1860s. The conference at St. Petersburg defined a kind of 'European culture' regarding the regulation of telecommunication that shaped the standardisation and governance of telecommunication infrastructures in Europe up to the late twentieth century. It was based on two elements:

(1) Cooperative standardisation within a committee was established as the sole way. The market played no longer a role for the standardisation of the telecommunication sector.

(2) International interconnectivity was made the major guideline for standardisation. Priority was put on low- or medium-level standards. Interoperable networks and high-level integrated networks were not envisioned. The national (sovereign) right to shield terminal equipment by incompatible national standards was not disputed.

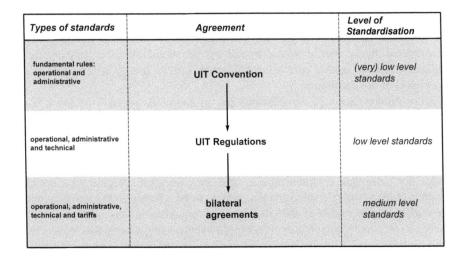

| Types of standards | Agreement | Level of Standardisation |
| --- | --- | --- |
| fundamental rules: operational and administrative | UIT Convention | (very) low level standards |
| operational, administrative and technical | UIT Regulations | low level standards |
| operational, administrative, technical and tariffs | bilateral agreements | medium level standards |

**Figure 7.1    Multi-level system of standardisation in the 1870s**

Henceforward, this pattern should shape the protagonists' thinking. In the following decades, new telecommunication infrastructures were grasped with the categories of the pattern. This had not been successfully challenged before the infrastructural policies of the European Union in the 1990s.

*3.3 The Regime Change in the 1920s*

In the decades that followed the conference of St. Petersburg the pattern of the 'European culture' concerning the regulation of telecommunication was taken into account when it came to the standardisation of telegraph networks. Low standards for operational and administrative aspects as well as for tariffs were negotiated within the UIT. The Regulations ensured the interconnectivity of the networks, whereas the basic principles within the Convention remained unchanged. National telegraph administrations negotiated the Telegraph Regulations rather independently at administrative conferences which were called regularly. The governments completely withdrew from the field of telegraph cooperation. UIT was subsequently transformed into an organisation of (national) administrations' technocratic experts who shaped the international relations rather independently. This was also the case because they frequently met, in contrast to technicians and engineers who did not.

The pattern for the integration of the telegraph was also adopted for the new modes of telecommunication afterwards: telephone and radio. When the telephone became an element of the telecommunication regime, the protagonists did not reflect on alternatives because they had internalised the pattern and the telephone was mostly put under the authority of the monopolistic telegraph administration.

The standardisation of different aspects of telephone services was carried out within the same institutional arrangements and in 1903 the telephone was included into the Convention and the Regulations for telegraphy.

The incorporation of radio into the regime took a more indirect way. Since the 1890s radio, which had a strong focus on maritime applications at the beginning, became subsequently integrated into the pattern of standardisation. However, in the early years of radio there had been a strong tendency to a deviant way of integration. The private Marconi Company tried to establish a global monopoly and therefore prohibited the intercommunication with other service providers. The idea was a forced competition that should eliminate all competitors. By doing so, Marconi violated the basic principles of the European culture for the regulation of telecommunication and provoked fierce opposition by the national PTT-administrations. It was the German government that decided then to call a preliminary conference in 1903. This conference should stop the activities of the Marconi Company and integrate the radio sector into the rules and standards for telecommunication.

Still before the Great War, the radio sector was subsequently adapted to the existing institutional arrangements for telecommunication and the fundamental design principles of the 'European culture' for the regulation of telecommunication were transferred to radio. A Convention was adopted which included – as the Telegraph Convention did – fundamental rules for radio communications. The Radio Regulations, annexed to the Convention, contained rules of technical, operational and administrative standards as well as juridical matters. However, the technical characteristics of radio forced to make some adjustments:

(1) Technical standards were pronounced compared with the Regulations for the telegraph and the telephone. Even the level of standardisation was higher.

(2) Radio was put in line with the 'European culture' for the regulation of telecommunication, but not incorporated into the institutions of the telecommunication regime. The International Radiotelegraph Union (Union Radiotélégraphique Internationale – UIR), which was founded according to the model of UIT, was instructed to carry out the periodic revision of the Convention and the Radio Regulations on either plenipotentiary or administrative conferences. Like in the telegraph sector, the national radio administrations and the industry negotiated the Radio Regulations rather independently (in the aftermath) at regular administrative conferences (Codding 1952).

The telecommunication regime hardly changed its character after 1875, meanwhile the telecommunication sectors expanded rapidly. New services were developed and telecommunication was transformed from an elite phenomenon into a modern mass media. Subsequently, different national development paths made equipment lose in technical compatibility, even though the UIT/UIR-standards still ensured the interconnectivity of national networks at the borderlines. The level of technical standardisation was lower than the one for administrative or operational standards at the turn of the century. These developments were to a large extent the result of national telecommunication markets which lacked competition. On the one hand, national PTT-administrations exercised a demand monopoly at several equipment markets (telegraph, telephone and radio) that enabled them to direct technical standards to the telecommunication industry. On the other hand, the PTT-administrations protected the national industry from foreign competitors by setting up trade barriers with the help of technically incompatible standards.

The changing telecommunication sector challenged then the multi-level standardisation system in the first two decades of the twentieth century.

(1) Telephone had reached a new stage of development in the early 1920s. Theoretically, international calls in higher quality and quantity were made possible. However, the technical differences of national telephone networks limited the potential of the system.

(2) Telephone and radio applications with a large number of private users needed further standardisations (bandwidth, transmission frequencies) in order to prevent national networks from drifting apart. A raise in the level of technical standardisation was urgently needed.

(3) The US-based General Electric tried to build up a pan-European telephone network that would have introduced high-level standards from outside.

(4) A transnational technocratic elite consisting of nation PTT experts had emerged due to the permanent cooperation within the UIT or the negotiations on bilateral agreements. Even informal cooperation became usual among these experts. They appreciated the benefits from cooperative standardisation in the fields of operation, administration and tariffs. An incorporation of technical standardisation into the telecommunication regime and an increase in the level of standardisation was looming. These four factors exerted strong pressure on the telecommunication regime to adjust its institutional arrangements to a new reality. An institutional spill-over' was approaching.

The decisive step was then taken by the French PTT-administration. In January 1924 it called for a conference in Paris in order to enlarge the telecommunication regime with a forum to negotiate technical standards (Laborie 2010). The PTT-administrations' technocratic elite across Europe supported such an idea, as they did not want the industry to standardise long distance telephony outside the telecommunication regime and the 'European culture' for telecommunication regulation. The representatives from Western European PTT-administrations agreed that the expanding European telephone system necessitated an institution which allowed telephone experts (from administrations and private companies) of various countries to convene periodically in order to exchange views on technical and operational issues concerning long-distance telephony. They founded the nongovernmental 'Consultative Committee for International Telephony' (Comité Consultatif International des Téléphonique – CCIF). The CCIF was in accord with the 'European culture' of telecommunication regulation.

It was a meeting place to prepare recommendations for technical and operational standards which did not violate national sovereignty. CCIF-recommendations were non-binding by law, even though they soon got a high commitment (quasi-standard). National telephone cartels between the PTT-administration and the industry were even strengthened. The PTT-administrations maintained their influence on technical standards and consequently extended their power in the international arena.

The CCIF was even provided with a permanent secretariat to prepare the annual meetings. The new organisation was established as an independent one, although a close connection to the UIT was self-evident. It was widely accepted that the committee could carry out valuable preparatory work for the revision of Telephone Regulations at the UIT's administrative conferences. At the following conference in Paris in 1925, the representatives from national administrations were already adding an article to the revised Telegraph Regulations which put the CCIF in charge of studying the technical and operational regulations (Laborie 2008). Although the representatives at the Paris conference were not endowed with the power to change the UIT's structure, they de facto did so. The 'technocratic automatism' of the integration is getting quite obvious here.

The 'institutional spill-over' was not limited to the telephone sector. It even spread across different modes of telecommunication. The CCIF's institutional design had already been transferred to the electric telegraph in 1925, when the International Telegraph Consultative Committee (Comité Consultatif International des Télégraphique – CCIT) was set up. Radio followed in 1927 when the Washington radio conference established the International Radio Consultative Committee (Comité Consultatif International des Radiocommunication – CCIR) which was similar to those for telegraphy and telephony (Wormbs 2011).

The setting up of the CCIs was an 'institutional spill-over' within the telecommunication regime which was fully in line with the norms and values of the 'European culture' of telecommunication regulation. The multi-level system of standardisation was complemented with the CCI standards and the average level of standardisation increased. The cooperative element was even strengthened because the CCIF established a 'permanent laboratory' in 1928. It should offer CCIF members an opportunity to cooperatively conduct research and studies. The importance of bilateral agreements for the standardisation decreased as a consequence of the CCIs.

The regime change was formally concluded at the Madrid telegraph conference in 1932 with the merger of the UIT and the UIR into the International Telecommunication Union (ITU). Two organisations with a similar structure were joined together in order to eliminate a duplication of work. The liaison was merely a juridical issue. Henceforward telegraph, telephone and radio operated under the provisions of a shared Convention, although the day-to-day work continued to be strictly divided (Codding 1952).

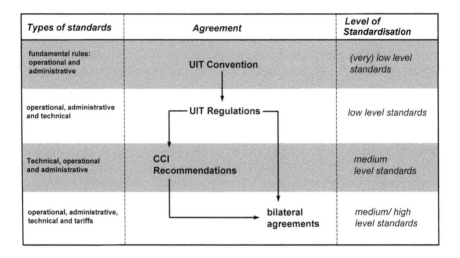

| Types of standards | Agreement | Level of Standardisation |
|---|---|---|
| fundamental rules: operational and administrative | UIT Convention | (very) low level standards |
| operational, administrative and technical | UIT Regulations | low level standards |
| Technical, operational and administrative | CCI Recommendations | medium level standards |
| operational, administrative, technical and tariffs | bilateral agreements | medium/ high level standards |

**Figure 7.2     Multi-level system of standardisation in the 1920s**

*3.4 The Regime Change in the 1950s and 1960s*

The telecommunication sector continued its expansion in the Interwar years and the decades following World War II. Terminal equipment like the telephone or broadcasting receiver entered the home and became part of everyday life. The average household was now connected to national and international communication spaces. Simultaneously, telecommunication services multiplied and challenged the standards-setting institutions. Especially the radio sector produced more and more services like broadcasting, air navigation, radio amateurs and from the late 1950s onwards even satellite communication.

The CCIs responded with the setting up of ever new study groups. Therefore, the functionalistic evolution of the telecommunication regimes continued. Technical research and technical standardisation increasingly became cooperative. Competing standards could hardly develop without recognition by potential competitors. Market standardisation was out of the question. The numerous contacts within the institutional arrangements accelerated knowledge transfers and interpersonal relations. The CCIs met very regularly and the UIT even had its own journal, the Telecommunication Journal. All types of standards were often negotiated within personal correspondence between the participating actors. It became more and more usual to forward concerns of technical research or standardisation within the CCIs. Foresighted standardisation and long-term periods of validity of the individual standards became ever more important. This was also the case as a change of technical development paths got more difficult with the number of users. The CCI-recommendations soon got a high commitment. They often turned into quasi-standards. This was also based on the incorporation of all relevant actors into the CCIs. The CCIs made a strong contribution to the establishment of intercompatible networks in Europe during the period of the internationalisation of telephone and radio (Henrich-Franke 2005).

Nevertheless, the CCI-recommendations obtained different levels of integration (Codding/Rutkowski 1982):

(1) They achieved the highest level of integration in the fields of wires that were necessary to interconnect networks. Here, we can observe a tendency towards unification in Europe. In contrast to the nineteenth century, it was no longer gateway technologies that separated distinct national telecommunication networks.

(2) Regarding terminal equipment the CCIs were most successful when the number of users and the corresponding economic interest of the industry were low. Then, recommendations reached a medium or high-level of standardisation. Meanwhile, as soon as the number of users and the corresponding economic interest of the industry increased, the cooperative standardisation within the CCIs often reached its limit. The PTT-administration's engineers often tried to establish interoperable telecommunication equipment, but they failed repeatedly. That markets of a high economic and political value were separated by technical incompatibilities corresponded with the 'European culture' of telecommunication regulation. Low-level standardisation and technological protectionism was disreputable. Therefore, it was the standardisation of lucrative markets for terminal equipment like television receivers where the CCIs reached their limits. Here, coalitions between politics and the economy overruled the PTT-administrations' technocrats (Fickers 2007). De Gaulle's intervention into the standardisation of colour television is the most prominent example. Here, functionalistic organisations reach their limits.

In the 1950s and the 1960s a new change of telecommunication regime was looming due to a number of indicators. (1) Technical innovations like satellite communication forced engineers to think in transnational categories. Neither could satellite beams be restricted to the small territories of European countries nor was it worthwhile for European states to invest individually into satellite technology. The imperative of national independency lost its importance in these cases. Now for the first time in the evolution of the telecommunication regime the European PTTs had to consider common undertakings and aspects of a shared telecommunication policy. (2) The ITU changed its character. With regard to the geographical scope of its activities, the ITU had changed considerably. What had started as a predominantly European organisation in the 1860s, had in the 1950s been transformed into a global one. The harmonisation and coordination of intra-European telecommunication infrastructures was just one aspect among others (Codding/Rutkowski 1982). Topics like tariffs for international telecommunication services in Europe or intra-European standards (especially in the very high frequency bands) could no longer be discussed solely within the CCIs. There, only the globally lowest common denominator was made a CCI-recommendation.

In a considerable number of aspects the CCIs recommendations decreased in the level of standardisation. This was for example the case in the standardisation of broadcasting equipment and transmission technologies, where a higher level of technological standardisation was prevented by developing countries that could not bear the costs of new technologies (Henrich-Franke 2005). Changes in the ITU called for a European organisation with a twofold aim: to discuss intra-European topics and to negotiate common European positions for global standard-setting activities. Especially the USA had turned out to be a dominating actor within the CCIs after its entry into UIT in 1932. Europeans were increasingly forced to agree on common European positions if they wanted to avoid to be marginalised.

The process of European integration was an external factor that accelerated the functionalistic development. European politicians aimed at founding a European PTT-organisation with strong connections to the institutions of European integration like the Council of Europe or the European Economic Community (EEC) since the beginning of the 1950s. These attempts met with stiff opposition from the PTT-administrations. They disliked the idea of giving up their traditional independency and even more disliked the EEC's idea of merging national telecommunication markets in a common interoperable European single market. This would have meant a violation of the basic principles of the 'European culture' regarding telecommunication regulation (Van Laer 2006).

The threat of a fundamental caesura in the telecommunication regime's development resulted in a 'technocratic automatism' that triggered off an 'institutional spill-over'. In 1959, the nongovernmental European Conference of Postal and Telecommunication Administrations (Conférence Européenne des Administrations des Postes et des Télécommunications – CEPT) was founded by 23 (Western) European PTT-administrations from 19 countries. This organisation was not linked to other projects of European integration, but rather incorporated into the existing telecommunication regime as a kind of sub-organisation of the ITU (Franke 2004). The CEPT was an institutional supplement which continued the traditional technocratic way of telecommunication governance in Europe. As a consequence of its nongovernmental inter-administrative structure, the CEPT was only authorised to make non-binding recommendations (Neutsch 2007).

The founding of the CEPT must be seen as an 'institutional spill-over' within the telecommunication regime that corresponded with the 'European culture' of telecommunication regulation. It was a European organisation to discuss intra-European topics and to negotiate common European positions for global standard-setting activities. It sponsored the foundation of the Conférence Européenne des Télécommunications par Satellites (CETS) among its members in 1963. The CETS coordinated European interests within the negotiations on a global (commercial) satellite network and carried out a European program for telecommunication satellite tests. The CEPT had successfully developed a model for common European undertakings by presenting the institutional design of the CETS. Also, in the field of administrative standardisations, the CEPT managed to establish a system of mutual clearings (CEPT Clearing) which had been impossible to realise within ITU. The CEPT Clearing was adjusted to the special conditions of an intensive intra-European information flow (Henrich-Franke 2014).

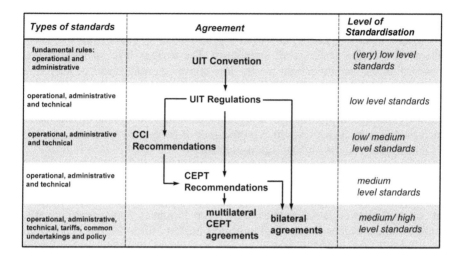

| Types of standards | Agreement | Level of Standardisation |
|---|---|---|
| fundamental rules: operational and administrative | UIT Convention | (very) low level standards |
| operational, administrative and technical | UIT Regulations | low level standards |
| operational, administrative and technical | CCI Recommendations | low/ medium level standards |
| operational, administrative and technical | CEPT Recommendations | medium level standards |
| operational, administrative, technical, tariffs, common undertakings and policy | multilateral CEPT agreements    bilateral agreements | medium/ high level standards |

**Figure 7.3    Multi-level system of standardisation in the 1960s**

*3.5 Towards a Regime Caesura*

Nevertheless, all the signs clearly showed that the 'European culture' of telecommunication regulation, which had been established in the nineteenth century, was endangered in its existence, even though the CEPT had temporarily postponed a caesura. The CEPT conflicted more and more with the European Community (EC) and a liberalising Zeitgeist due to the imperative of protecting national telecommunication markets by allowing technical incompatibility standards. From the 1980s onwards the EC broke up national PTT-monopolies and established the interoperability of infrastructural networks as imperative for standardisation within Europe. By doing so, the EC deprived the 'European culture' of telecommunication regulation of its fundaments. It even weakened cooperative standardisation in committees and strengthened standardisation by competition at open markets (Tegge 1992). This was a radical caesura of the telecommunication regime that had a different quality compared with the regime changes considered here (Noam 1992).

## 4. Conclusion

The aim of this chapter was to explain and compare pan-European regime changes in the telecommunication sectors in the nineteenth and twentieth century by making use of the functionalistic and neo-functionalistic thinking. We were able to observe different types of 'spill-over' that resulted in an increasingly integrated telecommunication sector in all three cases. The average level of international standardisation increased from a very low one in the nineteenth century to a medium or high-level standardisation in the twentieth century. More and more types of standards were negotiated in a cooperative way, while at the same time the regime expanded to all modes of telecommunication. Institutional arrangements increased in complexity. Overall, the processes of integration were characterised by an enormous continuity and path dependency.

The telecommunication regime's internal cohesion was guaranteed by a 'European culture' of telecommunication regulation which was made up of two components: the cooperative form of standardisation and the primacy of national network's interconnectivity over European interoperability. The right to shield domestic telecommunication markets from foreign competitors was never questioned. The durability and momentum of a telecommunication Weltanschauung becomes evident.

Increasing material interdependencies and the functionalistic logic of integration resulted in a revision of institutional arrangements which progressively approached the functionalists' ideal solution of an administration of things. The change of economic and technical demands accelerated the process of integration and put pressure on the institutional arrangements to change. The benefits from successful cooperative standardisation made the protagonists incorporate more and more types of standards into the telecommunication regime: in the 1870s it was tariffs, in the 1920s it were technical standards and in the 1960s it were common undertakings and policy. The 'institutional spill-over' was also a necessity because the precept of cooperation had eliminated market standardisation.

The components of the European culture limited the degree to which standardisation was allowed to integrate telecommunication networks. Low and medium-level standards were the maximum that could be achieved. In the nineteenth and the early twentieth century this did not prevent telecommunication networks from becoming increasingly integrated. However, when telecommunication services transformed into mass phenomena this prohibited the realisation of interoperable networks upon which pan-European equipment markets should have become possible.

Each regime change had its own characteristics which resulted from path dependencies within the telecommunication sector and the contemporary context:

(1) In the 1870s we have seen just a limited number of 'spill-over'. The electric telegraph was the only mode of telecommunication at that time and the complexity of technical systems was low. Nevertheless, in 1875 the conference

in St Petersburg fixed the 'European culture' of telecommunication regulation and laid the fundaments of 'technocratic automatisms' that would determine the developments for more than a century. The paradigms and guidelines of this regime were transferred to other telecommunication infrastructures.

(2) The regime change in the 1920s was probably the most complex one. Here, nearly all different types of 'spill-over' became noticeable. A number of factors culminated and put technical standardisation on the agenda. The change of telecommunication from an elite phenomenon to a mass phenomenon put strong pressure on the protagonists to standardise telephone and radio technology in order to avoid a technological drifting apart of national networks in Europe. The existing institutional arrangements needed a supplement that allowed for a cooperative standardisation of technology (and operations) among all actors without violating the fundamental principles of the telecommunication regime. The CCIs were a solution that passed recommendations with a high commitment. This was not least the case because of the cartels on the (national) telecommunication markets.

(3) The regime change of the 1960s was shaped by its institutional environment that approached a looming economic spill-over. European integration accelerated the developments and reversed the functionalistic succession of economic and institutional spill-over. The telecommunication regime was refined in a geographical dimension because the process of globalisation had transformed the ITU from a Eurocentric to a global organisation that no longer satisfied the particular needs of intra-European telecommunication markets. The CEPT was a result of an 'institutional spill-over' that re-established the conditions of a Eurocentric standardisation institution and that maintained a medium-level standardisation. At the same time it demonstrated the durability of the 'European culture' of telecommunication regulation. 'Technocratic automatisms' rejected external endeavours towards a regime caesura.

It is remarkable that the regime changes and the periodisation of the developments in the telecommunication regime differ considerably from the political history periodisation according to the World Wars and the changes in the general global system. This makes telecommunication a special case for comparisons. An important step towards an integrated telecommunication regime in the 1920s took place at a time when global politics were in a period of disintegration. The logic of functional integration seemed to be unaffected by the ups and downs of world politics. This observation even confirms a hypothesis of Johan Schot and Vincent Lagendijk. They claim that the origin of the 'technocratic internationalism' in the electricity and transport sectors would lie in the Interwar period (Lagendijk/Schot 2009).

We have seen that the functionalistic interpretation of the evolution of international relations might be a helpful tool to explain long-term developments and to compare individual development stages. Further research now has to focus on different types of spill-over across different infrastructure sectors. We have to raise the question, in how far the 'European culture' of telecommunication regulation can be discovered in other infrastructure sectors.

## References:

Ahr, B. and Benz A. and Tölle I. 2010. Der Einfluss politischer Akteure auf tarifäre Integrationsbestrebungen einzelner Infratruktursektoren im 19. Jahrhundert, in *Internationale Politik und Integration europäischer Infrastrukturen in Geschichte und Gegenwart*, edited by G. Ambrosius, et al. Baden-Baden: Nomos, 9–36.

Codding, G. 1952. *The International Telecommunication Union. An Experiment in International Cooperation*. Leiden: Brill.

Codding, G. and Rutkowski, A. 1982. *The International Telecommunication Union in a Changing World*, Washington, DC: Artech House.

Fickers, A. 2007. *'Politique de la grandeur' versus 'Made in Germany'. Politische Kulturgeschichte der Technik am Beispiel der PAL-SECAM-Kontroverse*, München: Oldenbourgh.

Giering, C. 1997. *Europa zwischen Zweckverband und Superstaat*, Bonn: Europa Union Verlag.

Haas, E.B. 1958. *The Uniting of Europe. Political, Social and Economic Forces, 1950–1957*. Stanford, CA: Stanford University Press.

Haas, E.B. 1975. *The Obsolescence of Regional Integration Theory*. Berkeley, CA: Institute of International Studies.

Franke, C. 2004. Das Post- und Fernmeldewesen im europäischen Integrationsprozess der 1950/60er Jahre. *European Journal of Integration History*, 10(2), 95–117.

Henrich-Franke, C. 2005. *Globale Regulierungsproblematiken in historischer Perspektive: Der Fall des Funkfrequenzspektrums 1945–1988*, Baden-Baden: Nomos.

Henrich-Franke, C. 2014. Regulating Intra-European Connections: telecommunications and European integration 1950–1970, in *The UN and European Construction: A Historical Perspective*, edited by L. Mechi et al. Cambridge: Cambridge Scholar Press.

Laborie, L. 2008. A Missing Link? Telecommunictions Networks and European Integration 1945–1970, in *Networking Europe. Transnational Infrastructures and the Shaping of Europe, 1850–2000*, edited by E. Van der Vleuten et al. London: Watson, 187–216.

Laborie, L. 2010. Georges Valensi: Europe calling?, in *Materilizing Europe*, edited by A. Badenoch et al. London: Palgrave, 198–201.

Mitrany, D. 1943. *A Working Peace System*. Chicago, IL: Quadrangle books.

Neutsch, C. 2007. Integration in den Bereichen Post und Telekommunikation nach dem Zweiten Weltkrieg bis zur EWG-Erweiterung 1973, in *Internationalismus und Europäische Integration im Vergleich*, edited by C. Henrich-Franke et al. Baden-Baden: Nomos, 113–32

Noam, E.M. 1992. *Telecommunications in Europe*. New York: Oxford University Press.

Reindl, J. 1993. *Der Deutsch-Österreichische Telegraphenverein und die Entwicklung des deutschen Telegraphenwesens 1850–1871*. Frankfurt am Main: Peter Lang.

Rosamund, B. 2000. *Theories of European Integration*. New York: St. Martin's Press.

Sandholtz, W. and Stone Sweet, A. 2005. *European Integration and Supranational Governance*. Oxford: Oxford University Press.

Schot, J. 2010. Transnational Infrastructures and European Integration. A Conceptual Exploration, in, *Les trajectoires de l'innovation technologique et la construction europeenne. Trends in Technological Innovation and the European Construction*, edited by C. Bouneau et al. Bruxelles: Peter Lang.

Schot, J. and Lagendijk, V. 2008. Technocratic Internationalism in the Interwar Years: Building Europe on motorways and electricity networks, in *Journal of Modern European History* 6(2), 196–217.

Tegge, A. 1992. *Die Internationale Telekommunikations-Union – Organisation und Funktion einer Weltorganisation im Wandel*. Baden-Baden: Nomos 1994

Van Laer, A. 2006. Liberalization or Europeanization? The EEC commission's Policy on Public Procurement in Information Technology and Telecommunications (1957–1984). *Journal of European Integration History*, 12(2), 107–30.

Wenzlhümer, R. 2010. Editorial – Telecommunication and Globalization in the Nineteenth Century. *Historical Social Research*, 35(1), 7–18.

Wobring, M. 2007. Die Integration der europäischen Telegraphie in der zweiten Hälfte des 19. Jahrhunderts, in *Internationalismus und Europäische Integration im Vergleich*, edited by C. Henrich-Franke et al. Baden-Baden: Nomos, 83–112.

Wormbs, N. 2011. Technology-dependent commons: The example of frequency spectrum for broadcasting in Europe in the 1920s. *International Journal of the Commons*, 5(1), 92–109.

## List of Acronyms:

| | |
|---|---|
| CEPT | Conférence Européenne des Administrations des Postes et des Télécommunications |
| CETS | Conférence Européenne des Télécommunications par Satellites |
| CCIF | Comité Consultatif International des Téléphonique |
| CCIR | Comité Consultatif International des Radiocommunication |
| CCIT | Comité Consultatif International des Télégraphique |
| DÖTV | Deutsch Österreichischer Telegraphen Verein |
| EC | European Communities |
| EEC | European Economic Community |
| ITU | International Telecommunication Union |

| PTT | Postal, Telegraph, Telephone |
| UIR | Union Internationale de Radiotélégraphe |
| UIT | Union Internationale de Télégraphe |

# Chapter 8

# Transportation Infrastructure Integration in East Africa in a Historical Context

Jacqueline Klopp and George Makajuma

## 1. Introduction

The problems of infrastructure integration in Africa are profoundly shaped by geography, politics, and history. A large part of the continent lay in tropical zones with populations concentrated near water and agriculture. The conference of Berlin in 1884 determined Africa's political borders, and the states that emerged in the 1960s are relatively recent creations, forming within and in resistance to colonial conquest. This has led to a large number of countries in Africa, 40 per cent of them small and land-locked. This 'sovereign fragmentation' has led to a situation where, 'for a given geographic area, the region has the highest number of countries, with each, on average, sharing borders with four neighbors' (Ndulu 2006: 215). These neighbours often have 'different trade and macro policy regimes', and differing political systems, languages and institutions stemming from different histories and colonial legacies. These constraints, along with poor infrastructure, difficult terrain, numerous border controls, roadblocks and weak governance overall mean many parts of the continent remain isolated from each other. Numerous studies show that median transport costs within Africa are estimated to be much higher than other world regions, reducing economic activity, connectivity and prosperity (Limao and Venables 2001, World Bank 2009).

To redress this situation, the African Union's New Partnership for Africa (NEPAD) and other regional organisations on the continent are pushing for more integration. Improvements in and greater linkage of transportation infrastructure are central to this broader economic and political integration agenda on the continent. In this chapter, we give a critical overview of East African transportation infrastructure integration as part of an economic integration agenda in Africa. We begin with a brief discussion about the earliest regional integration project through rail during the colonial period. This provides the historical context and an interesting base of comparison for contemporary times. We then examine the current project of constructing a highway link between Kenya's seaport of Mombasa via the Country's capital of Nairobi and onwards to Ethiopia's capital of Addis Ababa, part of a greater Trans-African highway. We look at the dynamics that drive and influence these transportation projects for integration and explore and highlight some of the inter-related issues around governance and financing, noting continuities and changes from the colonial to post-colonial periods.

## 2. Colonial Railways, Integration, and Infrastructural Legacy

Prior to the onset of European imperial conquest in the nineteenth century and subsequent colonial state formation, the area that now comprises Eastern Africa consisted mostly of numerous small political units including hierarchical kingdoms and highly decentralised and consultative systems of rule. Ethiopia, which maintained its status as a country during this time, was unique in the region. Trade routes linked different communities, and caravans crossed the land by foot and plied the Great Lake and the Indian Ocean by boat. European colonialists found these 'indigenous transportation systems wholly unsatisfactory' for their purposes causing their thoughts to turn towards railroads (Headrick 1981: 192).

The idea for the railway line in Eastern Africa thus began as part of imperial foreign policy and was linked to Britain's strategic interests in cementing control of Egypt and securing the source of the Nile from possible German incursion. Investors who saw business and trade opportunities were also key supporters of the East African railway project, which was initially to be undertaken by the Imperial British East Africa Company (IBEAC). Others argued that the rail and the colonial expansion would help end the slave trade based in Uganda. Thus, a combination of political, economic and social arguments and actors came together in a coalition for building the railway. Critics in Britain, however, called the project a 'lunatic express' and questioned the motives, the means and the cost. While the issue was highly contentious, the railway advocates prevailed, and after IBEAC succumbed to internal conflicts, the British government took over the project and the railway was built between 1896 and 1902 (Miller 1971). The rail project allowed the British government control of a very vast area, which in 1895 was declared the East African protectorate. The two projects – imperial conquest and the railroad – were inconceivable without each other.

The railway infrastructure that was built in East Africa at this time was successful because it was pushed by a powerful collection of British imperial interests. Unlike other railway projects that reflected nationalist projects of new countries as in Canada (Berton 1971), the East African rail project was a colonial project which meant benefits for local people were assumed as a spin off but hardly given a thought. The push to recover costs in the colonies meant rail lines were linked to importing finished goods and extracting resources and agricultural produce and getting them to the ports for shipment to Europe (Due, 1979, Luiz 2010). The complexity and difficulties of building the railway included the opposition in Britain, the high costs (£5.5 million between 1896 and 1901) and the challenges of traversing hostile local communities[1] and wildlife areas (Miller 1971). Nevertheless, after many lives lost and numerous complex negotiations, the railway was built (albeit using cheap materials)[2] and profoundly altered the future of the region.

The railway system became the basis for a new form of integration in an area that previously had more decentralised forms of government and its own complex networks of trade and inter-action, many of which were disrupted by British conquest and were actively discouraged by the form of the state that emerged out of it (Ambler 1988, Waller 1993). Further, the loss of land and displacement that followed colonial consolidation facilitated by the railway project led to a redistribution of wealth and power that continues to play a complex role in Kenyan politics and policy up to the present (Kanyinga, Lumumba, and Amanor 2008, Klopp 2000, Sorrenson 1968).[3] It also impacted the initial development patterns of the area particularly by clustering human settlements, embryonic urban centers, near railway stations. The colonial government in Kenya thereafter developed the road network linking the railway line with the large-scale farmlands inhabited by the white settlers.

## 3. The Post-colonial Period: The Shift to Roads

In 1963, when the new country of Kenya emerged the railway had already influenced development patterns profoundly. Most visibly, a great deal of development and settlement occurred around train stations and the rail line. The most dramatic was, of course, Nairobi. Nairobi, East Africa's largest metropolis and the capital of Kenya, began as a small Uganda Railway outpost roughly midway between the port of Mombasa and the port of Kisumu on Lake Victoria, and railway engineers devised the first highly segregationist 'plan' for the then tiny town (Anyamba 2008, 60). Kenya's other large cities – Mombasa, Nakuru and Kisumu – and many other important but smaller cities lay along the path of original Uganda Railway. Rail and the dominant political interests behind it also shaped Kenya's colonial road infrastructure; the network was largely developed to link white settler farms and areas to the railway. 'African' areas and hence production sites were neglected. However, this did not necessarily mean that

---

1   Some of the fiercest resistance was in the Nandi hills in Western Kenya where Koitalel Arap Samoei led a resistance movement for 11 years until he was tricked and murdered by British Col. Richard Meinertzhagen in 1905.

2   Due notes that, 'the railways were build cheaply with light rails – some no more than 35 pounds – and universally with a gauge less than the standard of Western Europe and North America' (1979: 375).

3   These ownership patterns to this day may have implications for transportation infrastructure by creating barriers to the expansion of rail over private lands or determining where large projects will be put to generate substantial opportunities for land value capture.

the areas away from the rail line were not productive. One study noted that by the 1960s many of the productive cash crop areas in Kenya were not along the rail line and that the link between economic development and railway construction was hardly straightforward (O'Connor 1965, 24). However, later developments of the road transport infrastructure in Kenya were heavily shaped by the railway line – the Sessional Paper No. 10 of 1965 lay greater emphasis for additional public spending on roads (and other developments) in high potential areas (Republic of Kenya, 1965), which consequently, worsened regional inequalities in provision of infrastructure as regions perceived as of low economic potential were generally ignored.

In the early independence period, Kenya, like other African countries, made attempts to restructure its economy and pan-African integration was part of this and actively promoted by the United Nations at the time. However, integration was in tension with the perceived need to build independent states and also the dynamics of localised political and economic interests. In the case of East Africa greater regional integration actually existed when the three countries, Kenya, Tanzania and Uganda, were ruled by one colonial power; prior to independence the three countries shared 'a customs union with a common external tariff and free trade between the countries, common customs and income tax administrations, common transport and communications services (railways, harbours, posts, telecommunications, airways), a common university, common research services, and a common currency' (Hazelwood 1979, 42). However, much of this infrastructure was skewed in favor of settler dominated and export-oriented economy with strong ties to Britain. African production and trade networks were sometimes facilitated during the colonial period by the infrastructure that brought regions together and sometimes actively discouraged for example, by laws restricting mobility and what kinds of crops could be grown and traded (Mugomba 1978).[4]

In 1967, Kenya joined Tanzania and Uganda to form the East African Community. Under this framework, the East African Railways Corporation (EARC), a well-managed public sector corporation, was the predominant carrier of freight traffic between Mombasa and Nairobi, and 'almost had a monopoly of long distance traffic into Uganda' (Republic of Kenya 2010, 20).[5] During this time, a number of tensions arose around regional transportation infrastructure. In particular, detailed provisions for managing a common road network were not included as part of the treaty that established the community. In Kenya, post-

---

4   African farmers, for example, were put at distinct disadvantages relative to white settler farmers. They could only travel with passes from the government, they were not allowed to grow certain crops and controls were put on commodities like maize and of course, they lost land which was the origin of the Mau Mau rebellion (Kanogo 1987).

5   Due notes that the East African Railways earned a small operating profit throughout the 1950s and 1960s after which deficits began to emerge-small at first but rising in the 1970s (1979, 382).

independence saw a rapid investment in roads and the expansion of private vehicles, especially those used to move petroleum products. Trade between Zambia and Kenya increased and heavy vehicles crossed Tanzanian roads imposing costs and straining relations (Hazelwood 1979, 48). Some argued also that the rise of private interests in transporting freight by road, helped explain the lack of needed investment in rail, although the politics over investment in transportation are most likely more complex (Hazelwood 1979, 48). However, the emerging disputes around transportation in the region did play into arguments over the distributions of benefits that would push the East African Community to fall apart by 1977 (Hazelwood 1979, Mugambo 1978). In the meantime, in parallel with the decline of the community, the East African Railways declined, a victim of competition from road transport, reduced efficiency, lagging rates and politics (Due, 1979: 384). More specifically, one report notes that KRC suffered from 'financial, technical and operational problems arising from poor corporate governance and inadequate investment, weaknesses in KRC management within a government-controlled environment and political interference in the appointment and tenure of senior management' (Republic of Kenya 2010).

## 4. Current State of Affairs

Following the collapse of the former East African Community (EAC) in 1977, the East African Railway network was broken up into three national railways: Kenya Railways Corporation (KRC), Tanzania Railways Corporation and Uganda Railways Corporation. In Kenya, entrepreneurship in transport has been encouraged in the post-colonial period, leading to many companies providing passenger and freight services albeit with poor regulation. It is, thus, unsurprising that with the formation of a strong road-based freight lobby emerging out of this entrepreneurship the dominance of road over rail in the post-independence period persists up to present.

The dominance of road over rail in the post-independence period is clear, even though this does not mean that Kenya's road system is adequate. Currently, the rail network consists of 2,778km comprising 1,083km of mainline, 346km of principle lines, 490km of minor and branch lines and 859km of private lines and sidings; over the last 10 years, only 38km of private line have been added (Republic of Kenya 2010). In contrast, according to the Kenya National Highway Authority, Kenya has a road network of about 177,800km with 63,575km classified. The classified road network increased from 41,800km at independence to 63,575km today. During the same period, the paved road length grew from 1,811km to 9,273km. It is presently estimated that about 70% (44,100km) of the classified road network is in good condition and is maintainable while the remaining 30% (18,900km) requires rehabilitation or reconstruction. Table 8.1 below gives a summary of classified road network in Kenya.

**Table 8.1    Classified Road Network**

| Road class | Premix | Length by Surface Type (km) | | | Total |
|---|---|---|---|---|---|
| | | *Surface dressing* | *Gravel* | *Earth* | |
| International Trunk Roads (A) | 1,244.91 | 1,563.81 | 715.11 | 94.48 | 3,618.31 |
| National Roads (B) | 350.21 | 1,166.26 | 819.29 | 346.14 | 2,681.90 |
| Primary Roads (C) | 642.89 | 2,198.16 | 3,601.64 | 1,552.90 | 7,995.59 |
| Secondary Roads (D) | 76.63 | 1,183.10 | 5,701.93 | 4,087.73 | 11,049.39 |
| Minor Roads (E) | 165.81 | 542.04 | 8,215.89 | 17,982.57 | 26,906.31 |
| Special Purpose Roads | 24.88 | 114.63 | 4,929.69 | 6,253.78 | 11,322.98 |
| All classes | 2,505.33 | 6,768 | 23,983.55 | 30,317.60 | 63,574.4 |

*Source:* Kenya National Highway Authority 2012 http://www.kenha.co.ke/index. php?option=com_content&view=article&id=46&Itemid=54.

It is worth noting that in the early 1980s rail's share of transport was at least 20% but this continued to decline until recently. Inadequate investments in maintenance and poor asset management also reflect a lack of political interest in rail from politicians who have invested heavily in private vehicles for public transportation and freight.

In the 1980s and 1990s, in line with predominant views around 'structural adjustment' and privatisation, multilateral donors such as the World Bank pushed commercialisation of the management and operations of KRC. This was unsuccessful, perhaps in part because of the lack of political interest in competition for roads-based transport, and rail services declined substantially.[6] A more recent concession agreement, which received support from the International Finance Corporation of the World Bank, leased the railways assets to Rift Valley Railways to manage and operate cargo transport services for 25 years and passenger services[7] for 5 years. This concession too has not led to improved performance

---

6    The cost of deferred maintenance and rehabilitation rises steadily over time when action is postponed.

7    Many poorer Kenyans rely on rail for transportation especially in the Nairobi area, but transportation for the majority does not appear to be a priority for the Kenyan government (Hagans 2011, Klopp 2011).

of rail transport, although more recently a number of big local investors with political connections including Equity Bank and Trans Century have infused more resources into the company (Michira and Okutah 2011).

Overall, the lack of focus and investment in rail represents a lost opportunity for an improved multi-modal and integrated transportation system for the region. Despite the problems, rail remains a main strategic inter-connector, linking Kenya and bordering countries to the largest port in East Africa – Mombasa. Besides Kenya and Uganda, the network currently also serves Eastern Democratic Republic of the Congo (DRC), Rwanda, Southern Sudan, and parts of Burundi, Congo, Somalia, Ethiopia and northern Tanzania (to some extent). Together these countries constitute a market of almost 200 million people. Over 80% of the cargo that passes through the port of Mombasa consists of imports, although this is set to change significantly as oil production in Uganda, Kenya and Sudan, copper mining in Uganda, and other mining opportunities in Eastern DRC build up. An estimated 8% of the port's throughput is currently transported by rail with the potential for expansion of this share (Republic of Kenya 2011).

Opportunity for deepening regional integration of railway networks is at present constrained by the relatively limited utilisation of existing lines (World Bank 2011). Rail traffic density in East Africa is the lowest on the continent and only a fraction of that found in southern Africa and North Africa, which are more developed. None of the East African railways (including the Tazara in Tanzania) has a traffic density in excess of one million traffic units per route kilometer. At the prevailing traffic volumes, it would be a mean feat for these organisations to generate enough revenue to finance track rehabilitation and upgrading, given their present deplorable state of affairs.

The benefits of a rail network go beyond trade facilitation. Rail possesses certain inherent advantages for long distance freight and passenger services over road transport, its main competitor (although of course, in a well-designed multi-modal system they should be complementary with a good road system that allows movement of goods and people to and from the rail lines). The benefits of such a system would include enhanced safety of goods and persons transported, security of cargo from theft, bulk hauling capacity, longevity of infrastructure, and lower maintenance costs, carbon emission and tariff per ton hauled (Coyle et al. 2011: 44–55). These attributes make rail the most economic solution for transporting bulk freight along major corridors over long distances.

From an economic standpoint, owing to its design, rail produces about twice as much transport capacity per dollar invested relative to a typical two-lane road. In addition, some bulk and container freight may not be suited for road haulage given road regulation and capacity constraints (Boyer 1997: 185–90). The Kenya–Uganda rail network, for example, has supported major infrastructure and mining operations in east and central African countries. These projects involve transportation of heavy machinery during the construction phase; and are dependent on safe and secure evacuation of bulk, sometimes hazardous, outputs to the international markets in the operation phase. Such freight could be completely captive to rail.

Regarding the Kenya–Uganda railway, several constraints have restricted operation of the network at full capacity. While the rolling stock and infrastructure are aging and for the most part, close to outliving their usefulness, the scarcity of financial resources need for huge long-term capital intensive projects implies that the assets will continue to perform poorly with regards to timeliness, reliability and efficiency (World Bank, 2010c: 113–14). Currently capacity utilisation for the rail track is at 40%, while locomotive utilisation is 75% of available fleet and wagons (African Development Bank 2011a). Derailments, more frequent in the recent past, are estimated to be as high as 800 per year and transit times between key origin-destination pairs have as much as doubled.

More recent attempts to revive the East African Community along with the problems with road congestion and lack of safety provide a new impetus to revive rail and road networks. On 30 November 1999, the Treaty for Establishment of the East African Community was signed and entered into force on 7 July 2000 following its ratification by the original three Partner States – Kenya, Uganda and Tanzania. The Republic of Rwanda and the Republic of Burundi joined the EAC Treaty on 18 June 2007 and became full Members of the Community as of 1 July 2007.[8] Article 89 of the Treaty states that member countries shall 'construct, maintain, upgrade, rehabilitate and integrate roads, railways, airports, pipelines and harbours in their territories' and 'review and re-design their intermodal transport systems and develop new routes within the Community for the transport of the type of goods and services produced in the Partner States'.[9]

The East African Community (EAC) project of economic integration is a key driver for the regional integration of transportation among the five states of Kenya, Uganda, Tanzania, Rwanda and Burundi. The overarching rationale articulated by proponents of this integration is to emulate a South-East Asia model where increased regional co-operation resulted in greater infrastructure investment and integration in the region, consequently leading to a virtuous cycle of investment, trade and co-operation (Asian Development Bank Institute 2008). EAC has suffered unfavourable terms of trade as it imports more goods and services than it exports in trade value terms (UNECA 2010). The fact that logistics costs, including transportation costs in East Africa, are higher than in most other regions in the world, including West Africa, which previously had the worst performance, plays into poor economic performance (World Bank 2011).[10] As an economic and

---

8   See http://www.eac.int/about-eac.html.

9   More recently the EAC hired a firm to produce an East African Railways master plan. See CPCS 2009.

10   There are, of course, many other problems. Poor communications connectivity and costly and unreliable power supply and distribution impose severe constraints on thousands of African enterprises (African Development Bank, 2011b). This further worsens the challenges due to inadequate physical infrastructure, slows transit speeds and increases

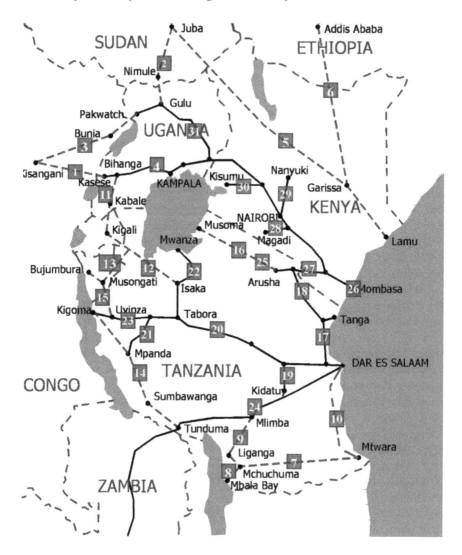

**Figure 8.1    Proposed East African Rail Network (CPCS 2009)**

political entity, the EAC ideal is increased competitiveness and ideally with more economic productivity and growth, improved standards of living.

---

costs imposed by weaknesses in trade facilitation, nontariff barriers, restrictive rules of origin not only for intra-African trade but also for major export markets in the rest of the world. A weak regulatory framework is another major undoing for private business and trade.

Unlike the past when integration was part of the colonial conquest, today a key player is the African Development Bank (AfDB), a multilateral institution with a majority of African country members. It has developed and is promoting a Regional Integration Strategy (RIS). In 2011, the AfDB provided a loan amounting to some USD 40 million towards the 5-year capital expenditure plan required under the restructuring and turnaround exercise of the Kenya–Uganda railway network, which is currently under a 25 year long concession to Rift Valley Railways (RVR). RVR has the potential to significantly increase the freight transported as a result of expanded capacity, faster throughput and improved reliability of the rail assets.

Key factors that can still derail these efforts include those leading to the difficulties already experienced with the Kenya–Uganda Concession. Since the deal was signed the lead equity investor failed to pay an initial capital commitment, further necessitating additional borrowing. Investments were subsequently delayed and the envisaged results failed to materialise in time for the populace. Public dissatisfaction with the performance of the RVR has increased political pressure to cancel the deal.

More recently, the Chinese government has signed an agreement to loan the Kenyan government $US 5.2 billion to fund the construction of a standard gauge rail line between Mombasa and Nairobi, the first leg of a rail line that is to eventually reach across the region. The China Roads and Bridges Corporation, which has entered an exclusive commercial contract with the Kenya Railway Corporation to construct the line, expects to finish by 2017. According to the company, the upgrade will cut passenger journey times from 12 hours to around 4 hours (Skynews 2013). However, this project has upset relations with its neighbors as it appears to undercut the existing agreement between the governments of Kenya and Uganda and the Rift Valley Railways, which holds a 25 year concession that could be violated by the Government of Kenya creating this new competitor (Africa Building 2013).

Overall, as this case shows external interests continue to play a heavy role in African infrastructure development, especially since few African governments and companies appear able to finance the large amounts of investment required to create an integrated transportation system. If we look at road construction and rehabilitation in Kenya, 56% of the finance between 2003 and 2008 came from external grants and loans from donors including prominently the World Bank, the African Development Bank, the EU and increasingly the Chinese government (Institute for Economic Affairs 2008). These donors work with the Ministries of Finance, Roads, Transport and Public Works to make decisions on transport planning, and they negotiate on contracts to do planning, feasibility studies, engineering design and construction. This leads to many problems including the failure to reap the maximum economic spin off effects of building infrastructure and the lack of accountability over transportation decisions that will have large impacts and incur substantial public debt. Competing foreign interests and weak governance are unlikely to produce optimal transportation decisions.

The upsurge of Chinese interest in infrastructure investment in the region could also complicate matters. While the Chinese government has had a long term engagement in African infrastructure, for example, helping to build the Tazara railway between Zambia and Tanzania in the 1970s (Monson 2009), recent Chinese investment in Africa and involvement in infrastructure development is unprecedented making China one of the largest donors on the continent (Brautigam 2009). This new dynamic is certainly evident in Kenya where the Chinese government is sometimes pitted against other external interests in transportation and integration projects. For example, in mid 2011 the Kenyan Government struggled to rebut an allegation that it was gradually warming up to Chinese financing and not the AfDB arranged financing in East African Rail. Recently, turning to Chinese support, Kenya did indeed pull out of a bilateral agreement to cooperate on an upgraded rail line from Mombasa to Kampala signed with Uganda in 2008 causing serious tensions and a possible lawsuit (Kisero 2011, Africa Building 2013). One attraction of Chinese financing for government officials is that it comes with few governance 'strings attached'. What is less clear is the actual cost and impact on public debt in the long-run due to less transparency involved in structuring such deals (The Economist, November 13, 2010). Overall, the lack of transparency in decision-making around transportation infrastructure more generally and the large extent of often competing foreign financing, means as in colonial times, a lack of downward accountability to citizens. As in the past, public debate is circumscribed (Klopp 2011) and overly narrow interests can prevail over sound policy around transportation needs and how best to approach regional integration.

## 5. The Trans-African Highway

In this section, we explore an attempt at integration of two very different African countries by road instead of rail. This case study illustrates that while trans-national roads seem to be favored and easier than rail to implement, like rail, integration via roads also faces many challenges. As in rail, some of the key actors in the financing, design and implementation of these projects are the Chinese government and multi-lateral institutions, particularly the African Development Bank and the World Bank. Both Banks emphasise the importance of individual African states expanding and improving their internal road network and also connecting them across countries, although the balance between focusing on internal and external connectivity and the extent of investment in road versus rail should be questions for policy debate but this rarely occurs.

As part of the broader integration strategy, regional bodies such as the East African Community and Inter Governmental Authority on Development (IGAD)[11]

---

11  The Intergovernmental Authority on Development (IGAD) in Eastern Africa was created in 1996 to supersede the Intergovernmental Authority on Drought and Development

supported by the African Development Bank are promoting the development of a transcontinental road network comprising seven major road corridors (African Development Bank 2003). Regional organisations and their individual member countries have been actively looking for funding since the 1990s for the implementation of a section of the Trans-African Highway Corridor from Cairo to Cape Town (UNECA 2010).[12]

This large road project includes a section that constitutes a major link between Nairobi and Ethiopian capital of Addis Ababa.[13] From Nairobi, the (Thika) highway goes through a large town called Thika and continues to traverse the northern part of Kenya from Isiolo to Moyale. Currently, according to an AfDB assessment, the main road connecting Addis Ababa to Nairobi has more than 700km of missing links including 366km of gravel road in Kenya and more than 300km of low standard and deteriorated paved road in Ethiopia (AfDB 2009, 2). Once the new road is completed by 2017 it is expected to remove these existing physical barriers to cross-border trade and help expand market size beyond national boundaries, although clearly many institutional barriers remain before better economic integration might occur.

In terms of achieving significant overall growth and enhancing magnitude and productivity of capital inflows as economies increasingly open up, infrastructure alone is clearly insufficient (Asian Development Bank Institute 2008). Infrastructure would have to complement the benefits of other incentives such as promotion of efficient financial intermediation, effective co-ordination of regional public goods, reduction of macroeconomic instability, and strengthening of security ties and effective institutions and hence improved governance.

As we noted, after independence Kenya substantially expanded its road network but in the later period, allowed this system like its railway to get run down. Since 2002 when Kenya elected a new government, the country has substantially worked on its road networks, although in a pattern that started in colonial times, this has not always been implemented equitably, strategically or carefully. The government has also implemented road sector reform focused on creating funds and better institutions and regulation to maintain roads. Currently, the Kenya Road Maintenance Fuel Levy Fund is the largest such fund in Sub-Saharan Africa. Similarly, in Ethiopia, after years of neglect and one of the lowest road densities on the continent, the government has been heavily

---

(IGADD), which was founded in 1986 to deal with recurring drought. Member states include Djibouti, Ethiopia, Kenya, Somalia, Sudan and Uganda. See http://igad.int/.

12   Interestingly, at the end of the nineteenth century the British businessman Cecil Rhodes had a similar vision for a Cape Town to Cairo rail and telegraph to facilitate conquest. The plan was initiated as a way to connect adjacent African possessions of the British Empire through a continuous line from Cape Town to Cairo.

13   The development of this road corridor had also been identified among the priorities of the African Union's New Partnership for African Development (NEPAD) Transport Program and was included in the infrastructure short-term action plan (2008–2012).

investing in road upgrading as in Kenya. In 2010, the Ethiopian Road Authority Director General claimed that the country's total road coverage was 49,000 kilometers and that this figure would rise to 136,000 kilometers, under a new plan, in five years (Tekle 2010). Like in Kenya the government has established the Ethiopian Road Fund to ensure that new roads are maintained.[14]

In part because of the neglect of rail, road carries more than 90% of motorised freight and passenger traffic within both of these countries, and the poor quality of roads imposes high costs on transportation of people and goods including in cross-border trade. Along with claims that roads lead to poverty reduction, this helps create the rationale for massive investment in roads in both countries. In addition, analysts from multilateral lending institutions, suggest that provision of an all-weather road connecting Ethiopia and Kenya is one of the best ways to foster bilateral and regional trade, economic growth and regional integration, the vision of the African Union. Current trade between the two countries has been largely by air given the poor condition of roads linking the two countries, and has remained marginal for many years (see Table 8.2). Ethiopia is also land locked and relies to some extent on Kenya's port in Mombasa. Currently, road conditions from Moyale to Mombasa are poor making exporting and importing of goods more expensive.

Ethiopia's level of bilateral trade flows is generally low with all its immediate neighbours. Ordinarily, being adjacent to a trading partner is expected to

**Table 8.2    Comparing Ethiopia's trade with immediate neighbours as of 2008**

| | Tariff KEN M | Tariff KEN X | Bilateral Distance | Partner GDP | 2008 two-way trade | 2008 two-way trade adjusted for tariffs | 2008 two-way trade adjusted for gravity |
|---|---|---|---|---|---|---|---|
| Kenya–Tanzania | 0.0% | 0.32% | 550 | 20,721 | 536.18 | 216.39 | |
| Kenya–Ethiopia | 1.45% | 16.32% | 1,186 | 25,658 | 33.40 | 33.90 | 124.20 |
| Kenya–Uganda | 0.0% | 0.27% | 479 | 14,529 | 586.75 | 221.38 | |

*Source:* Adopted from Ciuriak 2010.

---

14   It is worth noting that poorer countries tend to spend more on roads than richer countries but do not maintain them, ultimately costing them even more.

double the intensity of trade with that partner relative to non-adjacent countries (Frankel 2000). From the presentation in Table 8.2, actual two-way trade between Kenya and Ethiopia only amounted to US$ 33.4 million (about 6% of the level of two-way trade between Kenya and Tanzania). Kenya charges about 1.45% tariff on Ethiopian imports and, in turn, Ethiopia levies 16.32% tariff on Kenyan exports to Ethiopia. Some estimates suggest that the trade between Kenya and Ethiopia has the potential to increase substantially; however, this clearly depends on other factors including the very different politics and policies in both countries.

Based on the analysis of trade share between Ethiopia and its neighbours, one can quickly point a finger at the poor political relations in the region as a barrier to integration. Kenya, unlike Ethiopia, is a member of COMESA (Common Market for Eastern and Southern Africa) and has a more open economy (at least formally)[15] and much stronger industrial and services base than Ethiopia. Restrictions currently imposed by Ethiopia's neighbours hinges on the fact that the country has never subscribed to the Tripartite Free Trade Area (TFTA) or signed onto the COMESA Free Trade Agreement (FTA) which promises not only liberalisation of the goods trade but would also address the movement of persons, joint implementation of infrastructure projects and other forms of co-operation among the member states. Just as it has a more authoritarian government than Kenya, Ethiopia also has a much stricter set of controls over trade and reserves the right to control prices of basic commodities, which it did in 2010.

Recently, the Governments of Kenya and Ethiopia signed a Memorandum of Understanding to facilitate operations of transportation on this corridor. The two states are also working on transport service agreements/protocols on road transport, rail transport, maritime port facilities, and transit routes. However, this process has not been an easy one – it dragged on for at least two years fueled by Ethiopia's fears: Kenya has a much bigger market economy in contrast to the tightly controlled Ethiopian economy where even the National currency – the Ethiopian Birr, is heavily controlled and is not traded internationally. The Ethiopian government is wary that 'trade is in the favor of Kenya' which imports substantially less from Ethiopia than it exports (Nyabiage, 2011).

The other impediment has involved the different political systems and values. Ethiopia practices a fair amount of protectionism and is less tolerant to foreign ideologies (political, economic or social). Kenya, on the other hand, is struggling to become a democracy with a properly regulated market economy, which has seen it attract a fair amount of foreign direct investments relative to its neighbours. Kenya has continued to shape its business environment in a bid to encourage entrepreneurship and to enhance the ease of doing business for its fast growing population, although corruption remains a major problem. In Ethiopia, business enterprise is largely controlled by the state, which subsequently has stifled competition in the economy, and corruption, too is a problem. It has

---

15 Corruption remains a big issue in Kenya and can add an effective tax on some goods coming into the country especially at entry points like the Mombasa Port.

been the African Development Bank that played a fundamental role in navigating between these uneasy partners and ensuring that the transport service agreement and a Memorandum of Understanding between Kenya and Ethiopia was signed. This set a framework for promoting joint development and operation of transport, communications and other trade facilitation infrastructure, for enhancing trade, regional integration, cultural exchange and tourism between the two countries and the rest of the region. However, it is clear political challenges remain.

The two countries are now negotiating a simplified trade regime within COMESA that will list which items will be duty free but the lack of harmonised procedures for processing trans-border transactions persists as a problem. The AfDB also notes the need to harmonise transport regulation and implement effective transit operations on the corridor. The AfDB notes that, 'both countries ratified key COMESA instruments and common standards to facilitate regional transport and trade including harmonised axle load limits, harmonised transit charges, regional carrier licensing, regional third party motor insurance, Regional Custom Bonds guarantee and Singles Custom document' (AfDB 2009: 2). However, given the poor record of enforcement of transport regulations in Kenya, it is clear that issues of implementation will persist (Beja 2011, Kenya Anti-Corruption Commission 2007).

Other potential benefits/challenges of/to the Kenya–Ethiopia transportation corridor are linked to the history and geography of the border region between Kenya and Ethiopia. The highway project encompasses the arid northern part of Kenya and southern part of Ethiopia, where many pastoralists and subsistence farmers live. Water is scarce and possibly will become scarcer in the future, which means that the road construction which uses water and often contaminates it must be done very carefully and be properly regulated although realistically with weak governance systems this is unlikely. The AfDB is also arguing that the road will help improve access to markets and help transform the pastoral and subsistence agricultural environment into an area where small business can thrive, spurred by local and regional trade opportunities and increased income and access to goods. While, the entire area is underserved by transport and a road will lower transport costs, the road itself will not automatically produce affordable and accessible transport options, especially for the poor. Other supportive policy instruments will be needed to leverage the opportunities provided by the new road. Finally, the area has serious insecurity problems and by bringing in more actors and creating easier road access for security personnel, the road could contribute to the enhancement of security along the corridor. Once again these benefits of the road require complimentary policy and action that address problems arising from weak governance including in relation to security forces.

It is important to note that while the African Development Bank and the World Bank among other actors support regional and continental integration, ultimately national governments make decisions on transportation priorities (AfDB 2003, 21). However, as we have seen the politics of financing by external actors also shapes these decisions, as do internal power dynamics. Government actors can use the Trans-national Highway project as a way to bring in road projects based on different political and economic logics. In the case of Kenya, the 50.4km stretch from Nairobi to a small city called Thika, while being part of the trans-national highway, also became a way for some government actors to leverage AfDB funds for East Africa's first mega-highway that incidentally is raising land values in a region where many influential politicians own land (Namwaya 2004, Olingo 2013). Questions could be asked about whether a smaller, better designed highway up-grading scheme would have made more sense along with upgraded local rail line, rapid bus transit along the corridor and an improved regional road network around the capital. This might have better incorporated the interests of farmers, residents and businesses in the area and addressed congestion that exacts costs on the economy. This case also points to the need for improved integration of transportation projects in broader planning and public policy dialogues and greater context sensitivity in the actual design and building of infrastructure to maximally benefit citizens including those who do not own cars or are involved in long-distance freight by road (Beukes et al. 2011, Hagans 2011, Klopp 2011, Kara and CSUD 2012).[16]

## 6. Conclusions

While Africa has had very different historical experiences with integration issues compared to nineteenth and twentieth century European 'nation states', the historical experiences are intertwined. As European countries competed for control of African territory and resources, they also brought their world conflicts onto African soil, helping to create colonial states with undemocratic structures. African post-colonial states had to face these legacies of poor institutions, borders determined by European politics and also of infrastructure development built by external powers who largely had an extractive and external orientation. As we saw with the decline of the East African Community in the 1980s, the pan-Africanist vision of an integrated continent with integrated regions was quickly subverted to nationalist projects and the usual fractious politics of states. These politics were reflected in the actual infrastructure; as the East African Community declined so did the East African Railways Corporation and national road networks became the focus of concern. Today, the railway, a natural connector between Uganda and Kenya, which was born of imperial politics, now continues to be the subject of complex politics that still involves strong external influences linked to financing of infrastructure. This same politics influence highways and the relative balance of investment between these two transportation modes that ideally should be developed in a complimentary way into a locally adapted multi-modal

---

16  See the public debate around the Highway organised by the Kenya Alliance of Residents Associations at http://nairobiplanninginnovations.wordpress.com/2012/03/14/karacsud-public-forum-on-thika-highway-goes-online-2/ See also Kara and CSUD 2012.

transportation system. Such a system would produce better local, regional and continental mobility and access.

Organisations like the African Union, the African Development Bank and the numerous regional organisations continue to promote freer internal trade and movement of people on the continent and advocate and negotiate altered transportation infrastructure. The plans to rejuvenate rail networks in East Africa and to build the Trans-Africa Highway reflect this broader ideal in the East African context. Physical infrastructure like the Trans-African Highway has the potential to convert constraints of the natural geographic and historical challenges into economic opportunities by reducing costs, boosting productivity and raising the competitiveness of countries like Kenya and Ethiopia within a region and even globally, as well as enhance the flow of people and ideas. The complexities of the Ethiopia–Kenya case shows, however, building a highway between the two countries does little in the short term to deal with differences in history, politics and policy that mitigate against greater integration and interaction. Within many of these top down efforts there remains a whiff of the imperialist Lord Lugard, one of the key promoters of the Uganda Railway project: In 1922 he said that the 'the material development of Africa may be summed up by one word – transport' (Lugard 1922, Cited in O'Connor 1964: 21). Now, as then, we know that many other factors clearly come into play. East Africa, both in the past and through the current rejuvenated East African Community, which boasts a common passport, parliament and court, shows that such official integration of people across countries is possible when strong coalitions advocate for it.

Unsurprisingly, transportation infrastructure projects designed for better integration within Africa today continue to face many of the same political problems faced by Europe in cross-border collaboration and harmonisation, albeit made more severe by the large amount of 'sovereign fragmentation' and the weak institutions of many African states, which do not have supportive legal frameworks and clear and accountable ways to implement even those that exist.[17] Key institutional bottlenecks assume administrative, legal and regulatory form; current information asymmetries and discretionary powers in existing institutions encourage rent-seeking activities by state officials. Overall delays in customs clearance, unofficial payments, inefficiencies at the ports, and poor governance practices all create costs, including significant trade costs for companies and small traders alike (World Bank 2010). Issues around governance continue to increase costs on the roads and at borders but also on large-scale transportation projects (a phenomenon not unique to Africa). In the case of infrastructure, success will require a deeper break with the past than what we have witnessed in the last few decades of post-colonial history. By applying greater scrutiny of projects at the selection stage, integrity in procurement, efficiency in implementation, effective

---

17  For example, few have joined existing conventions and agreements such as the 1968 Convention and the 1949 Protocol on Roads signs and Signals (UNECA, AU and UN-OHRLLS 2011, 5).

post-completion management to ensure maintenance and efficient operation and continuing accountability to users is badly needed (Ndulu 2006). Fixing these problems is key part of the project of rebuilding more accountable and inclusive states on the continent, connecting people and economies and unleashing economic creativity and energy.

Overall, Africa has a large need for transportation infrastructure, which will require a great deal of financing. In continuity with the past, external actors including foreign governments and the World Bank are playing a key role. As in the past external financing is often linked to resource needs and geo-political strategies of donors. This politics of financing infrastructure, in interaction with governments that are not compelled to share information and analysis with citizens or have proper regulatory and accountability frameworks in place, can lead to skewed transportation and integration policies and projects. This dynamics, as in the past, can create new fragmentations rather than integration and facilitate skewed or unbalanced priorities where a more inclusive public interest is an afterthought[18] for as Divall and Revill note, 'transport throughout history as a practice heavily informed by, and informing, power' (2005: 101).

Overall, we have seen that the vision of an integrated Africa, well-connected through communications and transportation infrastructure, cultural interactions and trade, remains on the agenda. Part of this vision is in fact moving forward in regions like East Africa. We have seen that the challenges are many. In addition to the usual problems of harmonisation, Africans must deal with their relatively young states with problematic historical legacies that have shaped their institutions and relations to the wider world. In a context of declining availability of carbon-based fuels and growing concerns about climate change, Africans have the multiple struggles of national political change and battling significant poverty, while building badly needed transportation infrastructure, and re-integrating across often problematic borders, all this in a context of the continued strong influences of powerful external actors. With this in mind, the recent progress in East Africa towards improved transportation infrastructure and more integration can only be seen as significant achievement. Nevertheless, it is a small step on a longer, complex and contested journey towards transformation and possibly even integration.

---

18   For example, in the case of the Thika Highway, alternatives were not properly considered or discussed by Kenyan politicians and civil servants with the public and as a result, the actual highway may end up serving as a link in the trans-national highway but serving local citizens and businesses less well than it might (KARA and CSUD 2012).

## References

Africa Building. 2013. 'Kenya–China in secret Railway deal.' Accessed at http://www.africanbuilding.com/index.php/kenya-china-in-secret-railway-deal on December 12, 2013.

African Development Bank. 2003. 'Review of the Implementation Status of the Trans-African Highways and the Missing Links.' African Development Bank and UN Economic Commission for Africa (consultants: SWECO International AB. Sweden: Nordic Consulting Group AB, Sweden in association with BNETD, Ivory Coast and UNICONSULT, Kenya).

African Development Bank 2009. *Mombasa-Nairobi-Addis Ababa Road Corridor project Phase II: project Appraisal Report.* Unpublished report.

African Development Bank 2011a. *Investment Appraisal Model for the Kenya–Uganda Railways Concession Project, Private Sector Department.* Unpublished Report

African Development Bank 2011b. *Obstacles and Barriers to Regional Trade Integration in Africa*, Background Paper prepared by staff of the African Development Bank in collaboration with World Bank and WTO staff for the 2011 G20 Summit in France

Ambler, Charles. 1988. *Kenyan Communities in the Age of Imperialism.* New Haven, CT and London: Yale University Press.

Anastasiadou, Irene. 'Networks of Power: railway visions in inter-war Europe' *The Journal of Transport History* 28 (2): 172–91.

Anyamba, Tom. 2008. *'Diverse Informalities' . Spatial Transformations in Nairobi.* Saarbruken: VDM Verlag Dr. Miller.

Asian Development Bank Institute 2008. *Infrastructure and Trade in Asia*, edited by Douglas H. Brooks and Jayant Menon, Edward Elgar Publishing: 2–5.

Beja, Patrick. 2011 'Transporters demand cargo owners to be prosecuted for overloading' *East African Standard.* March 28. http://www.standardmedia.co.ke/archives/InsidePage.php?id=2000032091&cid=457&story=Transporters%20demand%20cargo.

Berton, Pierre. 1971. *The Last Spike: The Great Railway 1881–1885.* Random House of Canada.

Beukes, E.A., Vanderschuren, M.J.W.A. and Zuidgeest, M.H.P. 2011. Context sensitive multimodal road planning: a case study in Cape Town, South Africa. *Journal of transport geography*, 19 (3): 452–60.

Boyer D. Kenneth 1997. *Principles of Transportation Economics*, 1st Edition. Boston: Addison Wesley.

Brautigam, Deborah. 2009. *The Dragon's Gift: The Real Story of China in Africa.* Oxford: Oxford University Press.

Bullock, Richard 2009b. *Taking Stock of Railway Companies in Sub-Saharan Africa. Background Paper 17*, Africa Infrastructure Sector Diagnostic, World Bank, Washington, DC. East Africa Press, October 2009 available at http://

www.eastafricapress.net/october2009edition/index.php?option=com_content &task=view&id=35&Itemid=9.

Ciuriak Dan. 2010. Supply and Demand side Constraints as barriers for Ethiopian Exports – Policy Options, presentations at the 5th Annual National Private Sector Development (PSD) Conference, June 2010, Addis Ababa, Ethiopia. pp. 49–57.

Coyle J. John, R.A. Novak, B. Bibson and E.J. Bardi 2011. *Transportation: A supply Chain Perspective*, 7th Edition. South-Western Cengage Learning. pp. 44–7.

CPCS Transcom International Ltd. 2012. The East African Railways Master Plan. Unpublished document.

Divall, Colin and George Revill. 2005. 'Cultures of Transport: Representation, practice and technology.' *Journal of Transport History* 26 (1): 99–111.

Due, John. 1979. 'The Problems of Rail Transport in Tropical Africa.' *The Journal of Developing Areas*, 13(4): pp. 375–93

Frankel Jeffrey. 2000. Globalization of the Economy, Kennedy School of Government, Cambridge MA: Harvard University.

Hagans, Collin. 2001. 'Public Transport for Development – Development for Whom? Opportunities for Poverty Reduction and Risks of Splintering Urbanism in Nairobi'. MSc Thesis in Urban Development Planning submitted to the Development Planning Unit, University College London.

Hazelwood, Arthur. 1979. 'The End of the East African Community: What are the Lessons for Regional Integration Schemes?' *Journal of Common Market Studies* 18(1): 40–58.

Headrick, Daniel.R, 1981. *The Tools of Empire*. Chapter 14: African Transportation: Dreams and Realities. Oxford University Press: 192–203.

Institute for Economic Affairs. 2008. 'Infrastructure-Road and Rail Sector: Budget Performance 2003–08 and Emerging Policy Issues.' *The Budget Focus* 22, May.

Kanogo, Tabitha. 1987. *Squatters and the Roots of Mau Mau: 1905–1963*. Nairobi and London: East African Educational Publishers and James Currey Press.

Kara and CSUD. 2012. 'Thika Highway Improvement Project: A Social/ Community Co Analysis,' unpublished policy brief.

Karuti Kanyinga, Odenda Lumumba and Kojo Sebastian Amanor. 2008. The Struggle for Sustainable Land Management and Democratic Development in Kenya: A History of Greed and Grievance, in *Land and Sustainable Development in Africa*, edited by Kojo Sebastian Amanor and Sam Moyo. London: Zed Books.

Kenya Anti-Corruption Commission. 2007. Examination Report into the Systems, Policies, Procedures and Practices of the Roads Sub-Sector. Unpublished report.

Khayesi, M., Heiner Monheim, and Johannes Michael Nebe. 2010. Negotiating 'Streets for All' in Urban Transport Planning: The Case for Pedestrians, Cyclists and Street Vendors in Nairobi, Kenya'. *Antipode* 42(1): 103–26.

Kisero, Jaindi. 2011. 'China $4.6 billion railway deal to test Kenya's relations with Uganda', *The East African*, May 22, 2011.

Klopp, Jacqueline. 2000. "Pilfering the Public: The Problem of Land Grabbing in Contemporary Kenya", *Africa Today*, 47(1): 7–26.

Klopp, Jacqueline. 2011. 'Towards a Political Economy of Transportation in Nairobi'. *Urban Forum* DOI: 10.1007/s12132-011-9116-y.

Limao, Nuno and Venables, Anthony J. 2001. 'Infrastructure, Geographical Disadvantage and Transport Costs and Trade' *World Bank Economic Review* 15: 451–69.

Luiz, John. 2010. 'Infrastructure investment and its performance in Africa over the course of the twentieth century' *International Journal of Social Economics* 37 (7): 512–36.

Michira, Moses and Okutah. Mark. 2011. 'Kenya: Rift Valley Railway Secures Sh. 14.7 Billion Upgrade Debt'. *Business Daily*, 2 August. Available at http://www.propertykenya.com/news/1502314-cobrand!-.

Mijere, Nsolo. 2009. *Informal and Cross-Border Trade in the Southern African Development Community*. Ethiopia: OSSREA.

Miller, Charles. 1971. *The Lunatic Express*. New York: Macmillan.

Monson, Jamie. 2009 *Africa's Freedom Railway: How a Chinese Development Project Changed Lives and Livelihoods in Tanzania*. Bloomington: Indiana University Press.

Mugomba, Agrippah. 1978. Regional Organisations and African Underdevelopment: The Collapse of the East African Community. *The Journal of Modern African Studies* 16 (2): 261–72.

Namwaya, Otsieno. 2004. 'Who owns the land?' East African Standard October 1, 2004. Last accessed on December 18 2013 at http://www.marsgroupkenya.org/blog/2008/02/04/who-owns-kenya/.

Ndulu, Benno. Infrastructure, Regional Integration and Growth in Sub-Saharan Africa: Dealing with the disadvantages of Geography and Sovereign Fragmentation. *Journal of African Economies, Supplement 2* 2006, v. 15: 212–44.

Nyabiage, Jevans. 2011. 'Ethiopia on the spot over trade barriers'. *Daily Nation*. June 30. http://www.eastafricaforum.net/2011/07/02/ethiopia-on-the-spot-over-trade-barriers/

O'Connor A. M. 1965. 'New Railway Construction and the Pattern of Economic Development in East Africa'. *Transactions of the Institute of British Geographers*, 36: 21–30.

Olinga, Allan. 2013. 'Thika – "the next big thing" in property'. *Daily Nation*. September 18, 2013. Last accessed on December 18, 2013 at http://mobile.nation.co.ke/lifestyle/Thika-The-next-big-thing-in-property-/-/1950774/1997920/-/format/xhtml/-/r1v152z/-/index.html

Republic of Kenya. 1965. *Sessional Paper No. 10 of 1965; African Socialism and its Application to Planning in Kenya*. Nairobi: Government Printers.

Republic of Kenya. 2010. *Sessional Paper on National Integrated Transport Policy*. Nairobi: Government Printers.

Republic of Kenya. 2011. *Economic Survey 2011, Ministry of Planning and Vision 2030*. Nairobi: Government Printers.

Skynews. 2013. 'Kenya launches China-funded railway'. Accessed at http://www.skynews.com.au/world/article.aspx?id=929217 on December 11, 2013.

Sorrenson, M.P.K.1968. *Origins of European Settlement in Kenya*. Nairobi: Oxford University Press.

Tekle, Tesfa-Alem. 2010. 'Ethiopia to build 82,500 km road network'. *Sudan Tribune*. August 22. Available at http://www.sudantribune.com/Ethiopia-to-build-82–500-km-road,36043.

The Economist 2010, *Buying up the World: The Coming Wave of Chinese Takeovers*, November 13–19.

UNECA 2010. *Assessing Regional Integration in Africa IV – Enhancing Intra-African Trade,* United Nations Economic Commission for Africa, Addis Ababa, Ethiopia: 13–25

UNECA, AU and UN-OHRLLS. 2011. *Report of the Experts Group Meeting to Validate the Report on Regional Norms for the Trans-African Highway and the Draft Intergovernmental Agreement*. September 19–20, Addis Ababa.

Vail, L., 1977: Railway development and colonial underdevelopment: the Nyasaland case, in Palmer, R. and Parsons, N. (eds), *The Roots of Rural Poverty in Central and Southern Africa*. Berkeley, CA: University of California Press, 365–95.

Waller, Richard. 1993. Acceptees and Aliens: Kikuyu Settlement in Maasailand, in by Spear, T. and Waller, R. (eds), *Being Maasai*. London: James Currey: 266–57.

World Bank. 2009. *Transport Prices and Costs in Africa: A Review of the International Corridors*. Washington, DC: World Bank.

World Bank. 2010a. *Africa's Infrastructure: A Time for Transformation*. Washington, DC: World Bank.

World Bank. 2010b. *The Logistics Performance Index 2010,* in *Connecting to Compete 2010: Trade Logistics in the Global Economy*. Washington, DC: World Bank.

World Bank. 2010c. *Africa's Infrastructure: A Time for Transformation*. Washington, DC: World Bank: 113–14

World Bank 2011. *East Africa's Infrastructure: A Continental Perspective, Policy Research Working Paper No.5844*, Washington, DC: World Bank.

## Chapter 9

# From Liberalism to Liberalisation: International Electricity Governance in the Twentieth Century

Vincent Lagendijk

Behind the wall socket electrons move across borders with ease, empowering electronic machinery and appliances wherever and whenever needed. Electricity systems within the European realm are well-integrated both materially and institutionally. While electricity moves around Europe with relative ease, consumers sometimes experience some friction; travelling around Europe one encounters at least six different wall plugs. The need for adapters to different plugs might very well be the only friction left in Europe's electricity system.[1] The contrast with electricity itself could not be greater.

This integrated system has two dimensions. First, there is a material dimension. Electricity is a grid-based system, where the grid is used for transmission and distribution of electric power (Kaijser 2003, p. 155). A sufficient technological sophistication is necessary to build transmission lines and transformation stations, and to enable the flow of electrons across borders. This implies that equipment on both sides of the border need to be compatible with each other, directly or via an adapter. Technical standards need to be agreed upon, or interface technologies needed to be developed, in order to facilitate integration of electricity systems. Without it the system simply cannot not perform.

A second dimension is institutional; 'rules of the game' need to be established. These are partially incorporated in a legal framework, and in a set of more informal rules and principles which can be called the 'rules of the system'.[2] These needs to correspond with the goal and aims of the system at hand. These aims can range from primary ones like providing an uninterrupted electricity supply, to secondary ones like competition between electricity providers. These two dimensions are not constant; they evolve to lessen friction to more integration across borders. Friction

---

1 This *status quo* is official policy since 1990, as the costs of change and the risks during a lengthy transition to a universal system do not outweigh its benefits. See (CENELEC and Winckler 1994, p. 150).

2 'Rules of the game' is inspired by the work of North (North 1990). 'Rules of the system' is based on the thoughts of Tom Hughes, who argues that large technological systems are based on a technical core as well as on an organisational structure.

in this chapter refers to legislative, political and technological features hampering international electricity exchanges.

This chapter compares two periods; from the late nineteenth century to the Interwar period, with the post-World War II period. Both are subdivided into two eras. The first era comprises the early years of electricity, lasting until World War I. The second era – Interwar years – was characterised by a contested process of legislation. The second period concerns the post-World War II recovery and reconfiguration, which gives way to the fourth era starting approximately in 1975, which is marked by market liberalisation impulses from the European Commission.

The focus is on the governance of international electricity connections. According to Rosenau the advent of new technologies shrank social, economic, political and geographical spaces, but also create new problems. Governance answers to these challenges and is here defined as 'the activities of governments, but it also includes the many other channels through which "commands" flow in the form of goals framed, directives issues, and policies pursued' (Rosenau 1995, pp. 14, 16). Thomas Hughes has argued that all elements – technological ones and others – contribute to the system goal (Hughes 1987, p. 51). In the case of the electricity system not only technological aims (securing a cost-efficient electricity supply) but also socio-political aims are at work, as pointed out by Pfaffenberger (Pfaffenberger 1990, p. 364). He claims that political values are embedded in technologies; these values are part of a discourse which determines the strength of these values (Pfaffenberger 1992, pp. 382–4). Such a discourse is constructed by policy-makers as well as in part by engineers that build and govern the system.[3]

The chapter argues that three major governance shifts – namely, during the Interwar, post-World War II, and since the 1970s – were the result of changed thinking about system goals – both technological and socio-political ones. These shifts were directed at influencing international electricity exchanges. The main thrust for change came from different actors. The chapter continues in a chronological order, followed by a conclusion which compares the Interwar period with the post-World War II period, and ends pondering about possible future governance changes.

## 1. Before the War: From Liberalism to Regulation

The first power plants and transmission lines appeared during the last decade of the nineteenth century. Cross-border electricity transmission immediately took place between European countries, as there was little friction to integration on a limited scale. On the contrary even, there were at least four incentives to

---

3   Jeffrey Herf has pointed out that engineers often act as ideologues, and that they sometimes pursued ideological aims, even when these conflicted with engineering rationality itself. See (Herf 1984, pp. 152–4). This theme is further developed in (Lagendijk 2012).

go beyond national borders. First, shared resources like the Rhine also formed political boundaries, and developing the river thus implied sharing the benefits. This necessitates novel forms of international cooperation.

Second, most electrical utilities closely coordinated their activities with industrial consumers to ensure a good *load factor*. Load factor is the ratio of the average load to the maximum load, best visualised as a curve where peaks are at times of large electricity consumption, and valleys when demand is low. In short, the 'flatter' the curve, the better the load factor. The challenge for power station managers is finding a customer base leading to a constant load, even if this means meeting energy demands across national border (Hughes 1987, pp. 51–82, 71).

A third incentive for international cooperation concerned reaping the benefits of combining various types of electricity production. Each type of electricity generation has its own characteristics. Coal-fired plants are capable of providing a stable flow, but at relatively high and fluctuating running fuel costs. Hydro-power depends on the availability of water. This varies per season but has very little running costs after high initial investments. It is thus preferable to use hydro power instead of burning coal whenever possible. Striking a favourable balance between various resources – a good economic mix – can be profitable, even across borders.

Finally, security of supply was favourably affected by collaborating across borders. By integrating regional networks incidents like malfunctions in plants or the network can be tackled with more ease as there is more capacity available to cover for immediate shortages. This became increasingly important over the course of the twentieth century as the socio-economic importance of electricity substantially increased.

Cross-border flows of electricity are possible once certain conditions are met. One of these is technological. In the early years of the twentieth century electricity, 60–70 kilovolt (kV) lines were used which only had a limited economic range. The introduction of higher transmission capacities (110 and later 220kV) during the Interwar period, and up to 400 and 600kV after World War II) limited transport losses, and made it economically interesting to expand the geographical scale of cooperation. Standardisation was stimulated by emerging engineering organisations, pioneered by the International Electro-Technical Commission (IEC, 1906). Further international professional organisations sprang up between the wars, institutionalisation of knowledge and technology exchange occurred (International Electrotechnical Commission 1906). By the 1920s, most Western European networks had adopted a frequency of 50 Hz (Varaschin 1997, p. 142).

Another condition is regulatory; legislation and international agreements should not hamper the flow of electricity. Before World War I no international regulation existed for international electricity flows, and neither governments nor international organisations infringed with the governance of such flows. This left the initiative to private actors, mostly powerful alliances between manufacturers of electrical equipment on the one hand and banks on the other.

This would significantly change after World War I, and continue to transform ever more after World War II.

## 2. Interwar Electricity Regulation

After World War I changes took place in the electricity regime on both national and international levels. National authorities increasingly gained control over the electricity sector after World War I. For one, this was related to the changing economic tide. New legislation for international electricity flows were modelled after the overall economic protectionist measures. During the war various government bodies came to run specific economic sectors in order to overcome shortages. After the war engineers and policy-makers agreed that further rational utilisation of resources and organisation of the sector was necessary. This also helped to lower the 'additional costs' of foreign finance and to decrease external dependence in general. As a consequence the interconnection of networks and rationalisation remained priorities, but ever more within the boundaries of nations (Hausman et al. 2008, pp. 125, 127–9).

Therefore, in most European countries, authorities assumed some form of oversight over in- and outbound flows of electricity, and began to develop transmission networks and production capacity, in particular of hydroelectricity. This not only concerned import and export regulations and concession systems, but also included laws making watercourses 'national', and restricting their accessibility to foreign investors.[4]

Electricity was also regarded as an instrument for social advancement. National and regional authorities came to see electricity as a national public service (Bouneau et al. 2007, p. 35). National electricity laws aimed to expand production capacity, to interconnect regional electricity systems, and to encourage a wider distribution of electricity. Especially rural areas were electrified in the name of social and economic progress, and sometimes for electoral purposes (Varaschin 1997, p. 35; Coutard 2001).

While national authorities issued restrictive legislation, international attempts were made to provide conventions for electricity transmission across borders. The League of Nations (LoN, 1919) was the pioneer in this regard, especially its Organisation for Communications and Transit (OCT). It dealt with a variety of transport modes (road, rail, maritime, inland waterways, air), and aimed to standardise and provide international regulatory regimes through conventions.[5] Electricity matters came to be part of the OCT's work due to discussions about

---

4   A rather complete overview of European legislation is given in Siegel 1930.

5   See Schipper et al. 2010. Communications at that time was not equated to telecommunications, but to means of transport. Telecommunications was a minor part of League activities, but mostly reserved for other international organisations. For the League's activities concerning radio, see Fleury 1983.

railways.[6] While discussing the electrification of railways of international importance, the cross-border movement of electricty over overhead lines divided the members. The Italian delegation proposed that countries with large hydroelectricity resources should be responsible for the traction. Although the suggestion was rejected the OCT thought the issue should be further studied, as international electricity transmission concerned one aspect 'of the much wider [...] question of the value of international agreements in assisting to bring about a rational exploitation of power [...]' (League of Nations 1923a, Annex 7, p. 33).

This study materialised in a 1923 report. While underlining the novelty of international, the study provided two very general draft conventions. The first was an attempt to settle matters of international transmission and transit of electricity. In general, the study suggested, all measures and solutions should fit within the 'limits of national laws' (League of Nations 1923b, p. 5). This includes the construction of new lines and installations, either directly by states or concessionary companies. The choices made for new transmission lines should be technical considerations, and not political ones or national frontiers. Transit of electricity should be free of special dues, apart from charges for expenses made.

The second Convention concerned the development of hydroelectric power in international watercourses. It aimed to arrange the construction of power plants in rivers or lakes with two or more riparian states. The production of new plants should be arranged and divided between these states, and installations in the river should not hamper navigation.

Only a handful of states ratified the Conventions; four for the one on transmission in transit, and five for the one on development of hydraulic power on international watercourses (Mance 1946, pp. 148–50). This provoked a response from internationally-oriented electrical engineers, many of which argued for a *laisser-faire* regime for international electricity transmission. One prominent engineer claimed that international connections 'can never have any but a useful and beneficial effect from all point of view' (Landry 1926, p. 1117). From 1929 onwards engineers started to advocate a European network, giving this liberal regime a specific geographical dimension. At the same time, this allowed them to form an alliance with European-minded politicians favouring European economic and political unification (Lagendijk 2008, Chapter 3; Lagendijk 2012). Though many electrical engineers disagreed with constructing a European supergrid, most did favour a more liberal international regime within the European realm.

## 3. Post-World War II Electricity Regulation

Several elements re-emerged after World War II. The first was thinking in terms of 'Europe'. A post-war consensus emerged among electrical engineers who ran

---

6   Van Laak has argued that the railroad was 'the formative model of modern infrastructure'. See Laak 2004, p. 58.

the national systems. This implied that network operators continued to think in terms of European optimisation and rationalisation – a second continuity. This was further bolstered by the Marshall Plan. Post-war planners, and not just the Americans, clearly realised that electricity was quickly becoming an increasingly important contributor to economic growth.

While the Marshall Plan aimed for rapid European economic reconstruction, it promoted closer European cooperation as a means to do so. Part of the support for the electricity sector hence was earmarked to foster closer collaboration between European countries. A collaborative body for electricity eventually grew out of the Marshall Plan: the informal and engineering organisation the Union for the Coordination of the Production and Transmission of Electricity (UCPTE). It successfully plead for a relaxation of formal rules of inter-state electricity exchanges between 1951 and 1956, a process the UCPTE labelled as 'liberalisation' (Lagendijk 2011, p. 297). This governance shift thus pushed for more international exchanges, and was led by engineers.

Most of the resultant electricity exchanges were in service of the national electricity supply system. The shared safety measures were designed to isolate incidents locally, and preventing them from spreading to neighbouring regions and countries. Other forms of collaboration aimed to cover seasonal shortages (often exchanged for seasonal hydroelectricity surpluses) and coordination of power plant maintenance. With electricity increasingly permeating society, the (both economic and non-economic) costs of a power failure were becoming higher. UCPTE therefore made strong efforts to the secure electricity supply. This is illustrated by Table 9.1. Although the actual exchange of electricity across borders grew annually, the proportion in terms of national consumption was actually quite low and fluctuating yearly.

**Table 9.1    Electricity Exchange within the UCPTE,\* 1967–1985**

| YEAR | 1967 | 1970 | 1975 | 1980 | 1985 |
|---|---|---|---|---|---|
| Absolute growth in % | 5.6 | 6.1 | 5.6 | 7.0 | 7.7 |
| Relative to consumption (%) | 5.0 | 15.7 | -0.2 | 8.0 | 4.0 |

*Source:* UCPTE Annual Reports.
\* The UCPTE members in this period are Austria, Belgium, France, the Federal Republic of Germany, Italy, Luxemburg, and the Netherlands

While delegating powers to the UCPTE, national (and sub-national) authorities nevertheless kept control over the electricity sector. During the Interwar period governments stepped up as regulators, they now also increasingly became (co-)owners. The financial and political instability of the 1930s already seriously hurt the private actors in the electricity sector and fractured organisational structures.

In some countries – like Austria, France, Spain and Italy, and most Central and Eastern European countries – the patchwork of regional and local electricity companies was integrated into a publicly owned 'giant', and foreign elements were removed from the sector. This meant the end of the presence of the multinational electricity enterprises. In other countries governments directly stimulated electricity production and distribution (Hausman et al. 2008, pp. 223, 236–7).

During the first decades after World War II, most barriers to international electricity flows were removed by the engineer-dominated UCPTE. This process of liberalisation primarily served the (sub-)national systems, and fitted with the post-war strategy of national policy-makers. The necessary technological integration did however lead to a closely integrated European system, operating according to similar technical parameters. Optimising the national systems thus required international cooperation as well as the removal of friction to cross-border exchanges. It thus makes sense to speak of a liberal-national system of governance.

### 4. The 1970s: A New European Governance

Another impulse for electricity cooperation came from the European integration process that started with the European Coal and Steel Community (ECSC, 1951). In the negotiations resulting in the Treaties of Rome (1957), establishing the Common Market as well as the Atomic Energy Community, the need for a common energy policy for the Community was identified (CACEU 1955). Abundant and cheaper energy was seen as a cornerstone of economic progress. The existing common market for coal have to be extended to include oil, gas and electricity. These energy carriers would also have to compete with each other.

In the eventual 1957 Treaty electricity (as was gas) was excluded. Discussions within the preparatory committee on energy made clear that electricity was not a 'normal' commodity that could become part of the Common Market. The fact that electricity was network-bounded and its movement determined by laws of physics disqualified it from inclusion in the Common Market (CACEU 1955). Trading electricity is fundamentally different from regular goods; it cannot be efficiently stored and its movements are bound by laws of physics. In essence, electricity was never *traded* but *exchanged* – a transaction form resembling clearance.

The final report nevertheless identified the energy sector (including electricity) in need of Community action, in order to ensure a sufficient energy production. Progressively and gradually the various energy forms – like coal – should be included into the Common Market (Comité Intergouvernemental créé par la Conférence de Messine 1956, pp. 126–9). In subsequent years the European Commission (EC) indeed made several attempts to forge a common energy policy, and as a part of that, introducing competition in the sector. These attempts were mostly false starts until 1985 (Padgett 1992, p. 56). The main hurdle were the existing domestic structures, acting as inert and a strong countervailing forces (Eising and Jabko 2001, pp. 744, 748). National and sub-national government hold considerable stakes and interests in the energy sector as a whole. Authorities extract significant taxes from energy sales, and intervene in the sector to achieve broad social, economic and industrial aims. More recently they seek to fulfil (parts of) their environmental policy (Helm 2002, p. 174).

In 1985 the EC's Single European Act aimed to complete the Common Market by 1992, and electricity was a part of that effort (Matláry 1997). To make electricity part of the Common Market the institutional setting needed to change. A main aim would then be lower energy prices for consumers through competition. Competition would be enabled by breaking the monopoly of the sector, by liberalising the electricity system. This involved network access to third parties, and breaking up the integrated aspects of transmission and production of electricity. National networks, since World War I the near exclusive domain of domestic actors, would also be opened to foreign competitors.

Ending these national monopolies, the EC argued, would transform the sector into an open internal market, whereby electricity is produced on a competitive basis. Electricity should be subject to Community policy regarding environmental protection and energy in general (Commission of the European Communities 1988, p. 70). Differences of fiscal and financial preferences to utilities, as well as state support, had to be harmonised within the Community. The same applied to ownership structures of network ownership and national electricity pricing systems. The latter in particular should become more transparent (Commission of the European Communities 1988, pp. 71, 74). The main benefits were to be an increase in energy trade between member states, an enhanced security of supply, reduced energy costs, and a further rationalisation of the sector (Padgett 1992, p. 57).

This governance shift had serious consequences for international electricity transfers. Now rationalisation would not be limited to within nations but also *across* European borders, as national markets were opened to foreign and new players. This second phase of liberalisation was European in character.

This paradigm shift in how to govern infrastructural systems like electricity had to shifting visions on efficiency. Over the course of the 1970s several international actors – including the World Bank and the International Monetary Fund – began to see state-ran companies and strongly regulated industries as inefficient. Since the 1980s the most potential appeared to be in market-style competition and unbundling of transmission and production functions (Hausman et al. 2008, p. 263). Soon after various governments engaged in privatisation and deregulation 'experiments' across the globe, notably in Chile, the United Kingdom as well as in New Zealand and the United States. In addition, large energy consumers wanted to be more independent in choosing their energy suppliers (Gilbert et al. 1996; Eising 2002; Lagendijk 2011).

In addition to changing perspectives on system rationality, novel technological developments made these institutional adjustments possible. Electricity system became increasingly permeated with information technology devices, like sensors, computers, relays, and communication lines. Due to these developments, electricity could be better guided through the network and safeguarded the electricity supply to consumers, and allowed monitoring the system for irregularities and sharing of information as in the Supervisory Control and Data Acquisition system. On the basis of these real-time data, attributing a price to a certain amount of electricity has become a lot easier, as well as determining its 'path' through the network (Thue 2013: 213–38).

The existing institutional flag-bearer UCPTE did not immediately cede to the EC's plans. The UCPTE claimed that the industry itself had taken the initiative to liberalise cross-border electricity flows in the 1950s, which had put the sector ahead of other sectors (UCPTE 1988, p. 95; Lagendijk 2011). Yet there overarching aim had always been efficiency as well as making the system as robust as possible, by building in safety measures in order to ensure a reliable electricity supply. Yet finally giving in this organisation transformed itself and assumed two new primary tasks. First, it focused upon system security in the new situation by establishing technical rules for system operation related to interconnected synchronous operation in the membership area. Second, it held a close watch on system adequacy through supplying information to members, market players, and authorities (UCTE 2002, p. 27). In this way, it helped to change the institutional dimensions of the electricity system from security-oriented to a trade-centred within the European Union.

The time-table set in 2003 for further liberalisation implied that the electricity market should be open to all non-household consumers by July 1, 2004, and to all customers by July 1, 2007 (European Commission 2001, p. 11). Today, at least according to a UCPTE president, 'electricity has [...] assumed the characteristics of a branded commodity, to which it is now possible to assign a name, or even a colour', it being green, wind, or atomic electricity (UCTE 2000, p. 5). We only need the occasional adapter to actually tap into it.

## 5. Conclusions

How can we, finally, compare and contrast the eras before and after World War II? In sum, the main governance shifted from liberal to nationalistic after World War I, and from nationalistic to liberal-national after World War II, and eventually under influence of the EC, to liberal-European since the 1970s. This is schematically represented in Table 9.2. With these shifts, the geographical unit of optimisation changed, and a new dominant group of actors was able to change the system's aim.

During the liberal era, technological limits restrained electricity supply to the local level. Despite the limit economic range, electricity flowed across borders unhampered by a lack of regulation. Contrary to other infrastructural systems, electricity networks did not quite represent a Europe of nation-states until 1914. Multinational enterprises, and not national authorities, were the main actor in the electricity sector. Connections across borders were built by them, helped by an early standardisation effort.

**Table 9.2    Electricity Governance Eras**

|                      | Liberal          | Nationalistic                            | Liberal-National                                                      | Liberal-European                                  |
|----------------------|------------------|------------------------------------------|-----------------------------------------------------------------------|---------------------------------------------------|
| Period               | 1880s–1914       | 1914–1945                                | 1945–1970s                                                            | 1970s–today                                       |
| Geographical Scale   | Local, Regional  | National                                 | National and European                                                 | European                                          |
| Main Drivers         | Private Actors   | National and Regional Authorities        | National Authorities and Internationally-Minded Experts               | European Companies and Policy-makers              |

After World War I, when electricity transport over longer distances increasingly became possible, nationalistic policy creates barriers to international exchange of electricity in favour of the home economy. The dominant actors thus were national politicians. Several reasons can be named for this change. The wartime experience indicated control over energy resources was crucial for the war effort. This also led to further integration of regional systems and further rationalisation. At the same time, national authorities started to guard domestic energy resources against foreign elements. These new regulations left less room for multinational enterprises and restricted the international flow of electricity. In sum, the important socio-political aims inherent in the electricity system were nurturing national growth and remaining economic independent in terms of energy. This legislative 'friction' limited the possibilities of international exchanges.

This is the balance sheet of the period lasting until to World War II. The liberalism prior to the Great War, with some cross-border connections, was replaced by nationalism as domestic interests were prioritised. This development did not go uncontested. Against the nationalistic grain international organisations claimed a larger role. Both professional associations as well as intergovernmental organisations argued for more liberal electricity regime. Although their calls went largely unheard, a consensus emerged among engineers that a gradual construction of a European system was the way to go, and this should go accompanied with a liberal system. Engineers saw this as a type of European integration to overcome the 'friction' caused by nationally-inspired restrictions.

The national economy remained important in the liberal-national era after World War II, but now opened up in order to use European cooperation to optimise and secure the domestic electricity supply. In other words, the balance between national priorities and international cooperation altered. In post-war years an integrated Europe became a widely shared aim among policy-makers and experts alike. Stimulated through the Marshall Plan, an informal body was founded to facilitate international electricity cooperation between western European countries, the UCPTE. The expert-led UCPTE successfully lobbied against national restrictions to international electricity flows in the 1950s. Within the electricity sector a particular type of European integration was devised. Although the emerging national systems were technologically closely connected, cross-border connections were primarily used to optimise and secure the supply of national systems.

A second fundamental governance shift appears in the 1970s. A second vector of European integration, initiated in the 1950s, launched proposals for a deepening and widening of European cooperation also initially included electricity. This gained momentum as paradigms about the efficiency of state-controlled sectors changed, energy-intensive industries longed for more competition, and technological advances provided more detailed and real-time data to determine prices, and allowed for more control on flows within the system. The EC envisioned a common energy market where competition between producers and forms of energy would lead to more efficiency and lower prices. This opened up a new era of liberalisation, where efficiency on the European level was the main aim.

## 6. More Change to Come?

Like electricity, its governance, too, always seems to be in flux. More change is bound to come, as in recent years new pressures for further governance adjustments are increasing. Challenges include climate change, leading the EU to set sustainability and $CO_2$ emission reductions targets for 2020, and the renewed doubts about the viability of nuclear energy, due to the recent incidents in the Japanese nuclear power stations at Fukushima. These pressures provide incentives for a search for alternative forms of energy suppliers, both in terms of technology as well as geography.

In general, two trends can be observed. A first trend is the increased attention for decentralised electricity generation like solar panels, wind turbines, and combined heat and power applications. On the one hand local and small-scale units can have positive environmental benefits and help improve security of supply. On the other, however, such units are often unpredictable and volatile in terms of production, which can lead to overcapacity on local networks. In addition, the introduction of more decentralised generation is hampered by regulatory barriers (Pepermans et al. 2005). Governance changes are thus required.

The second trend concerns a number of plans, a number of proposal are put forward, resembling the Interwar super-grid idea. One example is Desertec that aims to harvest the Saharan sun with gigantic fields of photovoltaic cells. Related plans, like Airtricity, envision large wind farms in the North Sea (Clery 2010; Von Hirschhausen 2010). Both require a considerable strengthening of network connections and capacity, and would lead to a significant increase in cross-border flows. As approval procedures are slow, a policy response here, too, seems to be necessary. Whether and how this will translate into a new era of electricity governance remains an open question.

## References

Bouneau, C., Derdevet, M. and Percebois, J., 2007. *Les réseaux électriques au coeur de la civilisation industrielle*, Boulogne: Timée-Editions.

CENELEC and Winckler, R., 1994. *Electrotechnical Standardization in Europe: A Tool for the Internal Market*, Brussels: Cenelec.

Central Archives of the Council of the European Union, Brussels (CACEU), fonds 3: Negotiations for the Treaties institutionalising the EEC and EURATOM, File Nego 65: Comité Intergouvernemental créé par la Conférence de Messine, Commission de l'énergie classique. Projet de rapport, MAE 441 f/55 mvo, October 10, 1955.

Clery, D., 2010. Sending African Sunlight to Europe, Special Delivery. *Science*, 329(5993), pp. 782–3.

Comité Intergouvernemental créé par la Conférence de Messine, 1956. *Rapport des Chefs de Délégation aux Ministres des Affaires Etrangères*, Brussels: Secrétariat.

Commission of the European Communities, 1988. *The Internal Energy Market*, Brussels.

Coutard, O., 2001. Imaginaire et developpement des reseaux techniques: Les apport de l'histoire de l'électrification rurale en France et aux Etats-Unis. *Réseaux*, 5(109), pp. 76–94.

Eising, R., 2002. Policy Learning in Embedded Negotiations: Explaining EU Electricity Liberalization. *International Organization*, 56(1), pp. 85–120.

Eising, R. and Jabko, N., 2001. Moving Targets: National Interests and Electricity Liberalization in the European Union. *Comparative Political Studies*, 34(7), pp. 742–67.

European Commission, 2001. *Mededeling van de Commissie aan het Europees Parlement en de Raad: 'Europese energie-infrastructuur'. Voorstel voor en beschikking van het Europees Parlement en de Raad tot wijziging van Beschikking nr. 1254/96/E.G. tot opstelling van richtsnoeren voor trans-Europese netwerken in de energiesector.* Brussels.

Fleury, A., 1983. La Suisse et Radio Nations. In *The League of Nations in retrospect: Proceedings of the symposium*. pp. 196–220.

Gilbert, R.J., Kahn, E.P. and Newberry, D.M., 1996. Introduction: International Comparisons of Electricity Regulation. In R.J. Gilbert and E.P. Kahn, eds, *International Comparisons of Electricity Regulation*. Cambridge: Cambridge University Press, pp. 1–24.

Hausman, W., Wilkins, M. and Hertner, P., 2008. *Global Electrification: Multinational Enterprise and International Finance in the History of Light and Power*. Cambridge: Cambridge University Press.

Helm, D., 2002. Energy Policy: Security of Supply, Sustainability and Competition. *Energy Policy*, 30(3), pp. 173–84.

Herf, J., 1984. *Reactionary Modernism: Technology, Culture, and Politics in Weimar and the Third Reich*. Cambridge: Cambridge University Press.

Von Hirschhausen, C., 2010. Developing a Supergrid. In B. Moselle, J. Padilla, and R. Schmalensee, eds, *Harnessing Renewable Energy in Electric Power Systems: Theory, Practice, Policy*. Washington, DC: Earthscan, pp. 181–206.

Hughes, T.P., 1987. The Evolution of Large Technical Systems. In W.E. Bijker, T.P. Hughes, and T.J. Pinch, eds, *The Social Construction of Technological Systems*. Cambridge, MA: MIT Press, pp. 51–82.

International Electrotechnical Commission, 1906. *Report of Preliminary Meeting*, London.

Kaijser, A., 2003. Redirecting Infrasystems Towards Sustainabilty. In *Individual and Structural Determinants of Environmental Practice*. Aldershot: Ashgate, pp. 152–79.

Laak, D. van, 2004. Technological Infrastructure, Concepts and Consequences. *ICON*, 10, pp. 53–64.

Lagendijk, V., 2008. *Electrifying Europe: The Power of Europe in the Construction of Electricity Networks*, Amsterdam: Aksant.

Lagendijk, V., 2011. 'An Experience Forgotten Today': Examining Two Rounds of European Electricity Liberalization. *History and Technology*, 27(3), pp. 291–310.

Lagendijk, V., 2012. 'To Consolidate Peace'? The International Electro-Technical Community and the Grid for the United States of Europe. *Journal of Contemporary History*, 47(2).

Landry, P., 1926. Exchange of Electrical Energy Between Countries: General Report on Section B. In *Transactions of the World Power Conference, Basle sectional meeting*. Basle: E. Birkhäuser and Cie., pp. 1112–24.

League of Nations 1923a. *Advisory and Technical Committee for Communications and Transit, Procès-verbal of the second session, held at Geneva, March 29th–31st, 1922*. LoN doc. Ser. C.378.M.171.1923.VIII (Geneva: League of Nations), Annex 7.

League of Nations 1923b. *Second General Conference on Communications and Transit. Electric Questions: Report concerning the draft Conventions and Statutes Relating to the Transmission in Transit of Electric Power and the Development of Hydraulic Power on Watercourses Forming Part of a Basin Situated in the Territory of Several States. Vol 3*. LoN doc. Ser. C.378.M.171.1923.VIII (Geneva: League of Nations).

Mance, S.O., 1946. *International Road Transport, Postal, Electricity and Miscellaneous Questions*, London: Oxford University Press.

Matláry, J.H., 1997. *Energy Policy in the European Union*, London: Palgrave Macmillan.

North, D.C., 1990. *Institutions, Institutional Change, and Economic Performance*, Cambridge: Cambridge University Press.

Padgett, S., 1992. The Single European Energy Market: The Politics of Realization. *Journal of Common Market Studies*, 30(1), pp. 53–75.

Pepermans, G. et al., 2005. Distributed generation: definition, benefits and issues. *Energy Policy*, 33(6), pp. 787–98.

Pfaffenberger, B., 1990. The Harsh Facts of Hydraulics: Technology and Society in Sri Lanka's Colonization Schemes. *Technology and Culture*, 31(3), pp. 361–97.

Pfaffenberger, B., 1992. Technological Dramas. *Science, Technology and Human Values*, 17(3), pp. 282–312.

Rosenau, J.N., 1995. Governance in the Twenty-First Century. *Global Governance*, 1(1), pp. 13–44.

Schipper, F., Lagendijk, V. and Anastasiadou, I., 2010. New Connections for an Old Continent: Rail, Road and Electricity in the League of Nations' Organisation for Communications and Transit. In A. Badenoch and A. Fickers, eds, *Europe Materializing? Transnational Infrastructures and the Project of Europe*. Houndsmills: Palgrave Macmillan, pp. 113–43.

Siegel, G., 1930. *Die Elektrizitätsgesetzgebung der Kulturländer de Erde*, Berlin: VDI-Verlag.

Thue, L., 2013. Connections, Criticality, and Complexity: Norwegian Electricity in its European Context. In *Europe Goes Critical: The Emergence and Governance of Transnational Infrastructure Vulnerabilities*. Cambridge, MA: MIT Press, pp. 213–38.

UCPTE, 1988. *Rapport annuel 1986–1987*. Heidelberg: UCPTE.

UCTE, 2000. *Rapport annuel 1999*. Brussels: UCTE.

UCTE, 2002. *Rapport annuel 2001*. Brussels: UCTE.

Varaschin, D., 1997. *Etats et électricité en Europe occidentale. Habilitation à diriger des recherches*. Habilitation. Université Pierre-Mendes-France: Grenoble III.

Chapter 10

# 'Wings for Peace' versus 'Airopia': Contested Visions of Post-war European Aviation in World War II Britain

S. Waqar Zaidi

## 1. Introduction

This chapter examines two internationalist and yet very different visions for post-war aviation which came to prominence in London during World War II. The first emerged from within the Labour and Liberal parties, and other internationalist organisations. It envisaged the integration and unification of European commercial aviation through a newly created pan-European airline. This airline was to be run by a transnational aviation authority created, managed and owned by European nation-states. An alternative vision imagined that this pan-European airline would be run by a private corporation, rather than an international authority answering to national governments. There were attempts to turn both visions into reality. In this chapter, I explore the first through discussions within the international relations think-tank Chatham House and Labour party policy statements. I show that this internationalist vision emerged from earlier interwar thinking on aviation, and was strongly evident in Britain in 1943 and 1944. Initially, proponents of this vision battled supporters of nationalised aviation. By 1944, however, a new vision of internationalised aviation, based on private ownership, came to the fore. I examine this vision through the attempts by the aerial entrepreneur Count de Lengerke and his supporters to materialise a pan-European 'Airopia' airline. I show that, in the end, the triumph of the American view of international aviation, combined with the nationalisation of aviation across Europe, led to the demise of both visions.

There have been numerous historical works investigating wartime thinking on post-war international aviation. This literature however has tended to see wartime thinking as a singular debate between varying hues of internationalists on the one hand and nationalists (that is those wanting international aviation to remain in the hands of national governments as opposed to supranational organisations) on the other (Brewin 1982, Devereux 1991, Dobson 1991, Van Vleck 2007, Kranakis 2010). This chapter changes this current dichotomous picture of nationalist versus internationalist thinking in several important ways. First, I show that there were not only varying hues of aerial internationalist proposals, but in fact opposing internationalist views which differed from each other in crucial ways. These opposing internationalist views of aviation battled each other as much as they battled nationalist conceptions. This chapter highlights the political and intellectual contexts within which these two opposed visions gestated and gained currency. I focus in particular on ideas and debates relating to the benefits of international organisation, the international nature of civil aviation, the needs of post-war European recovery and integration, and the role of private corporations in international transport. Here, I draw and build upon the work of Peter Lyth on the British government's attempt to direct British post-war aviation (Lyth 1995), on Robert Millward's studies of growing nationalisation of transport from 1945 onwards (Millward 2005), and Peter Wilson's work on British internationalist thinking during World War II (Wilson 1996).

Second, I focus on the rhetorical content of these visions and show that they made use of metaphors and public images of international companies and international transport dating back to the early interwar period. For internationalists, explaining their visions and convincing policymakers was not only about focusing on new technology and the purported future of international travel, but also about drawing upon metaphors based on long-existing systems of international travel. Internationalists made the argument for efficiency and rational organisation of post-war aviation by utilising powerful pre-existing widespread images of internationalised, interconnected and seamless pan-European rail travel. In this sense, this chapter builds on the recent work on internationalist projects for international transport and communications networks, both pan-European and global, in the interwar period and during World War II (Anastasiadou 2007, Schipper 2008, Zaidi 2011).


## 2. Wagons-Lits at Chatham House

Thinking about the post-war world began almost as soon as the war started, and planning in earnest not long after. Although much of this thinking was carried out in government, organisations outside of the state, such as the London-based international relations think-tank Chatham House, also became centres of thinking about the post-war period. Chatham House had an interest in the future of aviation and its relationship to international relations dating back to the 1920s. This interest continued into the war through the formation of a high-powered discussion group on aviation which met at least seven times between March and October 1943. Participants included aviation industry representatives such as Frederick Handley Page and R.H. Thornton (Chairman of the Aviation Committee of the General Council of British Shipping), leading aviation journalist Peter Masefield (of *The Aeroplane* and the *Sunday Times*), various government representatives (including Labour MP Philip Noel-Baker from the Ministry of War Transport), interested internationalists, (most prominently Air Chief-Marshal Sir Arthur Longmore and Jonathan Griffin, Author of *World Airways*), and Chatham House transportation experts such as Harry Osborne Mance and J.E. Wheeler.

Debate quickly coalesced around two positions on the future of international aviation. Aviation industry representatives emphasised the need for 'freedom of the skies', by which they meant the freedom for private aviation to fly across countries without being impeded by burdensome government regulation and control. Handley Page, for example, believed that it would be best to 'dissociate governments from the direct running of air lines and also from directly subsidising air lines'.[1] Internationalists, on the other hand, argued for the need for concerted inter-governmental control of international civil aviation. They argued that an international centrally organised pan-European aviation system would be the most efficient way of organising pan-European aviation. For Wheeler 'in a continent like Europe it was a great waste of money to have 25 different companies with 25 or more types of equipment'.[2] Only such international control would ensure that this (in Noel-Baker's words) 'most marvellous invention of modern times' would bring 'adventure, romance, happiness, health and understanding to the human race'.[3]

Such internationalisation had a pedigree stretching back to the late 1920s. Since then, British and French internationalists had been calling for the organisation of European aviation on a transnational basis. Responding to what they claimed was the inefficiency of European commercial aviation, and the threat of destruction posed by massive rival national air forces, internationalists had argued that both military and commercial European aviation should be controlled by the League of Nations or some such transnational authority. All aeroplanes or aviation facilities were to be either owned outright by a transnational authority, or would be tightly regulated by the authority which would allocate routes to national and private airlines (Zaidi 2011).

These proposals died out with worsening international relations across Europe in the late thirties, but were reborn with the onset of war. There was a sense in these newly rejuvenated arguments that a particular opportunity had arrived that would allow for aerial internationalisation to an extent which earlier had not been possible, and may not be possible again. The era of post-war European reconstruction opened up the possibility of reconstruction of, and reconstruction based around, aviation. Claims that political integration could be furthered through international co-operation in technical areas, particularly transport and communication, began once again to find purchase within internationalist and even the wider discourse (Wilson 1996). A significant manifestation of this was the publication of David Mitrany's Chatham House pamphlet *A Working Peace System*, wherein Mitrany claimed that aviation 'could be organised effectively only on a universal scale' (Mitrany 1943). European integration became widely

---

1   Chatham House. 24 March 1943. *Informal Discussion Meeting: The Future of Aviation*, 8/986, Chatham House Archives.

2   Chatham House. 19 May 1943. *Private Meeting: British Air Policy in the Pacific*, 8/987, Chatham House Archives.

3   Chatham House. 24 March 1943. *Informal discussion meeting: The Future of Aviation*, 8/986, Chatham House Archives.

talked about again in internationalist circles. Even the international relations expert E.H. Carr, an earlier critic of liberal internationalist thinking, called for formation of a European Reconstruction Corporation and a European Planning Authority as a first step towards the infrastructural and eventual political integration of Europe (Carr 1942).

In making the argument for a reconstructed internationalised aviation, internationalists pointed to what they saw as pre-existing successful examples of international co-operation. Arthur Longmore, a recently-retired senior Royal Air Force (RAF) officer, envisaged that Europe would undergo a transitional period of reconstruction which would require an internationally planned aviation. European aviation would need to be run by the 'air forces of the United Nations' who he noted, were 'already trained to operate together'. During this period, air forces of the lesser European powers could be organised into 'regional international air police forces' which would keep peace and security through air power. This aerial organisation would, in the short term, ensure the 'future peace of Europe'.[4] In making these arguments Longmore was drawing on a widespread belief, current at that time, that the Allies' air forces, led by the RAF, had demonstrated an advanced form of international co-operation which could serve as a model for future international organisation (Zaidi 2011).

Longmore however believed that something even more international and durable would be required for the longer term peace and stability of Europe. Here he suggested as a model an existing internationalised transport system co-habiting with nation-states and their national transport systems: that of the prominent international travel company Compagnie Internationale des Wagons-Lits et des Grands Express Européens, or more simply, Wagons-Lits.[5] Formed in 1876 and head-quartered in Paris, Wagons-Lits owned and operated a large number of luxury sleeping cars, some 1,600 at the eve of World War I. These attached onto national rail services across Europe, allowing passengers to travel in their first class sleeper carriages from one end of Europe to the other. By 1926, the Company claimed to transport two million passengers a year in this manner. The Company also operated express services, such as the Orient Express, and provided travel agency services through offices located in railway stations. This breadth of international operations gave Wagons-Lits its public reputation as the most international of European companies.[6] Nor unsurprisingly, the company did its utmost to project this image, and to suggest that Wagons-Lits promoted international peace. At a 1925 board-meeting the Chairman, in front of senior government officials and diplomats, boasted that:

---

4  Chatham House. 24 March 1943. *Informal Discussion Meeting: The Future of Aviation*, 8/986, Chatham House Archives.

5  Ibid.

6  For histories see: Behrend 1962, Behrend 1959.

The Wagons-Lits Company was now probably one of the most complex and far-reaching organizations in the world ... No greater factor existed for the interchange of good fellowship and good understanding than that which was afforded by the Wagons-Lits Company.[7]

He went on to claim that the 'Wagons-Lits Company carried a banner of international goodwill into every remote corner of Europe. In a modest way the board in itself was a small League of Nations'.[8]

By the end of World War I, due in part to nationalisation of its rolling-stock and facilities by Germany in 1914 and by Russia after the 1917 revolutions, Wagons-Lits was a greatly diminished player in international rail travel. This demise however did not prevent Wagons-Lits from occupying a prominent position in aerial discourse from the twenties onwards. In 1926 it was suggested in the House of Commons that 'civil aviation in all countries should be handed over to the League of Nations, which should form a sort of Société Internationale des Wagons-Lits de l'Air'.[9] In a 1928 House of Commons debate on aerial disarmament, a Labour MP called for commercial aviation to be 'put under international control, under the supervision of the League of Nations'. He continued:

There is nothing impracticable in the proposal that these great air services should be put into the hands of holding companies, in which the nationals of the different great air States would have their agreed proportion of stock, companies whose reports would be submitted to the League of Nations. If that were done, the air traveller of the future would proceed across Europe very much like the railway traveller does to-day under the International Wagons-Lits Company, with its headquarters at Brussels and with an international staff, which goes right across the frontier of Europe and takes you even as far as Asia Minor.[10]

These references to Wagons-Lits in proposals for the international organisation of aviation continued into the 1930s, and spread also to French internationalists (for example: W. Arnold-Forster 1931, Pierre Brossolette 1931). Opponents of internationalised aviation used Wagons-Lits in their statements too – they made the point that internationally owned civil aeroplanes could be seized by aggressive nations, just as Wagons-Lits carriages had been.[11] Wagons-Lits was also sometimes used more generally to denote luxurious air travel: the terms 'aerial Wagon-Lits' and 'wagon-lits of the air' were widely in use by 1929 (Parker 1928, Harper 1929).

---

7 International Good Will. Work of the Wagons-Lits Company. *The Times*, 20 June 1925.

8 Ibid.

9 HC Deb 08 March 1926 volume 192 column 1984.

10 HC Deb 12 March 1928 volume 214 column 1646.

11 For example in Parliament: HC Deb 29 November 1933 volume 283 column 972; and HL Deb 11 March 1937 volume 104 column 647.

Aviation enthusiasts looked forward to the eventual supplanting of international rail travel by aviation – Douhet noting that in the future 'the Compagnie des Wagons-Lits will have changed its name, and international mail will be carried by air transport' (Douhet 1942).

Like earlier proponents of the international organisation of aviation, Longmore too drew upon this pre-existing metaphor of Wagons-Lits during the Chatham House discussions. He envisaged intra-country aviation services continuing to operate alongside a new international aviation company with a monopoly on international European aviation. This was a realistic possibility, he believed, because the transitional nature of aviation in Europe in the immediate post-war period (aerial networks, airlines, and facilities were being recreated or built up for the first time) provided a unique opportunity for the rational reorganisation of international aviation. Also, the Allies would anyhow be heavily regulating international military and civil aviation immediately after the war, and aviation was known to be at the forefront of Allied planning for new post-war international organisations.[12]

Although the Chatham House discussions did not lead to any coherent policy recommendations, they never-the-less influenced the wider discourse on post-war aviation. Prompted by the discussions within this group, for example, *The Times* produced two feature articles on the 'Problems of Civil Aviation'. It noted that Europe was entering an 'experimental period in air travel' that called for further 'international organization of air transport'. This was to be done through the formation of an international 'regulatory authority' and an international airline, a 'Wagons-Lits of the air' connecting cities from Moscow to London.[13]

In August 1943, Arthur Longmore was asked to join a Foreign Office committee on post-war reconstruction as the RAF representative. The Post-Hostilities Planning Committee discussed, amongst other matters, the future of European aviation. Here too Longmore called for the control of international aviation through an international organisation, including the formation of an international air force (Longmore 1946).

A research paper by Mance, suggesting international control of aviation, and used as the initial basis for the Chatham House discussions, went on to be greeted with positive interest in the Foreign Office (Mance 1942).[14] Indeed, astonishingly, the British government itself, between late 1941 and March 1943, contemplated post-war international control of civil aviation as official policy

---

12   Chatham House. 19 May 1943. *Private Meeting: British Air Policy in the Pacific*, 8/987, Chatham House Archives. Also: Chatham House. 24 March 1943. *Informal discussion meeting: The Future of Aviation*, 8/986, Chatham House Archives. For more on wartime planning see Zaidi (2009).

13   1943. Problems of Civil Aviation: II – Danger of International Rivalries, The Special Case of Europe. *The Times*, 30 December.

14   Foreign Office Filenote. 5 January 1943. *Future of International Civil Aviation*, FO 371/36430, National Archives, Kew.

(Brewin 1982, Devereux 1991, Zaidi 2011). A (largely Liberal) constituency within the civil service and government argued that it provided an expeditious way of protecting international British commercial aviation from the expected post-war international expansion of American airlines, and of extending British aviation across Europe. The choice was, as one report put it, between 'Americanisation' or 'internationalisation'.[15] The Cabinet, meeting in June 1943, expressed disappointment that British proposals for complete internationalised control of international civil aviation were not being enthusiastically greeted by the Dominions and the Americans. Foreign Secretary Anthony Eden expressed the hope that not-withstanding the lack of success for global proposals, aviation in Europe could never-the-less be internationalised. Winston Churchill not only agreed but emphasised that this internationalised civil aviation should not simply be an 'amalgamation' of different countries' aviation interests, but rather should be controlled through an international pan-European organisation. Wagons-Lits, he announced to the Cabinet, was the model to be followed.[16]

### 3. Labour Party, Reaction, and Government

The Labour Party, home of many internationalists, was also a locus for proposals for the internationalisation of aviation. In April 1944 the Party's policy paper *Wings for Peace* demanded that a 'World Air Authority' own and operate a 'World Airways' with a monopoly on international trunk routes. The Authority would own *all* civil aircraft, even those leased to national airlines servicing domestic routes. The paper conceded that such internationalisation would not occur immediately: the interim aim would be internationalisation on a regional basis throughout the world, including the formation of a pan-European 'Europa Airways' and a 'British Commonwealth Airways'. This would allow for the growth of aviation internationally at lower cost and with less wasteful competition, unifying the peoples of the world and bringing peace (The Labour Party 1944).

The most vocal of Labour supporters was the MP Frank Bowles who formed a World Airways Joint Committee which propagandised for a 'World Communications Authority' which would start with aviation but eventually include 'all international communications – railways, shipping, telegraphs, posts etc.' (Bowles 1944, World Airways Joint Committee 1944, Bowles and Vernon 1946). This Committee was funded by the prominent wartime internationalist organisation the World Unity Movement. Michael Dunlop Young, Director of PEP (1941–1945) and (from 1945) head of Labour research, wrote press articles and pamphlets calling for internationalisation throughout the war (Young 1944, Joyce and Young 1945).

---

15   Lord Finlay. 18 December 1942. *Internationalisation of Civil Aviation After the War*, War Cabinet Report RP(42) 48, AIR 2/5491, National Archives, Kew.

16   Cabinet Secretary. 24 June 1943. W.M.(43), 88th Meeting (24 June 1943), Cabinet Secretaries' Notebooks on War Cabinet Meetings, CAB 195/2, National Archives, Kew.

As late as 1947 in a significant policy document he emphasised Labour's continued commitment to world government alongside internationalised civil aviation (Young 1947).

Press opinion on the internationalisation of civil aviation split in much the same way it had done in the thirties. From the Centre-Left the *Manchester Guardian* declared *Wings for Peace* a 'bold framework' for international aviation, and supported regionalised internationalisation, which it felt would be more acceptable to the Americans.[17] Opponents, such as *The Times*, retorted that private competition was required for development of aviation, and that anyhow the Labour Party proposals were unrealistic as they did not take American opposition into account.[18] The Conservative Party organised opposition through the formation of a lobbying committee of four Conservative MPs (one Vice-President of the British Airline Pilots Association). Their manifesto reasoned that internationalisation was not required for 'air security', 'would not be acceptable either to the USA or to the USSR' and was not 'practical politics' (Tree 1943). *The Economist*'s response fell in between that of the rejectionists and proponents. Although it conceded that 'complete international control' was 'the ideal solution', it, like *The Times*, criticised *Wings for Peace* for ignoring Great Power objections. It suggested a compromise between this ideal solution and 'freedom of the skies': 'Freedom of the International Air, with the Commonwealth treated as a unit, and with a mutual sharing of imperial cabotage rights with the French and Dutch Empires.'[19]

A proposal for complete internationalisation, akin to the Labour proposals under discussion in Britain over the past year, was tabled by the Australian and the New Zealand delegations at the November 1944 International Civil Aviation Conference in Chicago. The American delegation however insisted on an 'open skies' policy, and this vision eventually won. Other countries were forced to agree: airline routes and flying would henceforth be agreed on a bilateral basis between nation-states (Zaidi 2011).

## 4. Airopia

The Labourite vision of international control of aviation died following the November 1944 conference. At the same time, however, an alternate vision of internationalised aviation arose in wartime Britain. This shared, and indeed attempted to materialise, the Wagons-Lits model. This alternate vision was the brainchild of George de Lengerke, but quickly won backing amongst Conservatives

---

17    1944. World Aviation, *Manchester Guardian,* 18 October; 1943. Air Transport, *Manchester Guardian,* 13 March.

18    1944. A World Air Authority, *The Times,* 27 April.

19    1943. Air Transport, *The Economist* 144, 6 March, 286–8. 1943. Civil Aviation, *The Economist* 144, 5 June, 713–14;1944. Civil Aviation, *The Economist* 146, 1 April, 424–6; 1944. Wings for Peace, *The Economist* 146, 29 April, 583.

and the political right. It envisaged the creation of a private pan-European airline, Airopia, which would run air services in Europe, and have its shareholdings spread across Europe. The vision was internationalist, though without state control – described by one civil servant as 'a skilful blending of internationalisation and private profit'.[20]

Airopia Ltd was formed in London in March 1944 by George Edward De Lengerke, who took a major shareholding and appointed himself Managing Director. Lengerke was born in London on 30 April 1895, and educated at the Engineering School of Lusanne University. He served in the (British) Secret Service during World War I and then in the twenties (or perhaps the 1930s) helped establish the Imperial Airways route to the Far East through Italy. He also at that time served as Secretary of the Italian airline S.A.N.A. In November 1939 he fled Budapest for London. Lengerke's main supporters were the wealthy businessmen Major John R. Chambers (Chairman of Airopia's Board of Directors) and Director Captain Leonard Frank Plugge. Chambers was then Deputy Director of Clothing and Textiles at the Ministry of Supply. Conservative MP Leonard Plugge was a pioneer in commercial radio broadcasting during the interwar years, and had chaired the Parliamentary and Scientific Committee during the war. He had served with the RAF during World War I, and was particularly well connected with several aircraft manufacturers (Wallis 2007). Air Commodore Sir Adrian

**Figure 10.1    Airopia logo, dated July 1945**

*Source:* From: AIR 2/2508, National Archives, Kew. Reproduced with permission of the National Archives.

---

20    Riddoch, J.H. to E.B. Bowyer. 26 July 1944. *Memorandum*, AIR 2/2508, National Archives, Kew.

Chamier, onetime Secretary General of the Air League of the British Empire, was also a Director.

The initial plans were ambitious, though they would dwindle down over the years due to British government opposition. The May 1944 *The Lengerke Plan for the Development of Civil Aviation after the War* presented itself as a compromise between entirely state-controlled international aviation and unfettered international private competition. Lengerke envisaged two aerial regimes: one for flying within and between European countries, and the other for flying between and within countries outside of Europe (including European colonial possessions and the Commonwealth). Outside of Europe aviation was to be organised by an 'Airport Corporation' which would have 'sole authority to regulate air traffic'. It would allocate and regulate international routes, and own outright all ground facilities. National Governments, alongside 'private enterprise', were to take shareholdings in proportion to the aviation facilities they handed over to the Corporation. Private and nationally-owned airlines would be allowed to operate on these routes, subject to the Corporation's approval and regulation. For Europe, he suggested the formation of a separate 'Inter-European Airport Corporation' which, like its global counterpart, would be owned by national governments and private interests in proportion to the facilities they hand over to the Corporation. Just like for the non-European case, a multitude of companies would be allowed to service routes within European countries. But the crucial difference to the non-European situation was that only one company was to have the monopoly on international travel within Europe: Lengerke's airline, Airopia. The shareholdings of the company was to be divided between Lengerke and his partners, and between other private shareholders whose holdings would be restricted to countries served by the airline, and in fixed proportion. Great Britain, France, and the USSR were to receive twice as many shares as Holland, Belgium, Italy and Germany. The Directorships were also to be allocated amongst countries on a similar basis.[21] Lengerke believed that 800 aircraft would ultimately be required to complete his network across Europe. Initially however he hoped to purchase 25 York aircraft which would form the nucleus of his airline. It was to be based initially in London, but he planned to move it to Switzerland soon after the end of the war. Services were to start from London via Zurich to Istanbul, and Lisbon via Madrid and Zurich to Stockholm.[22] The use of a single airline such as Airopia to fly Europe's air routes, as opposed to the situation before the war, would lead to significant cost savings – 30 per cent to 40 per cent cheaper than before the war, he estimated. Like government-orientated internationalist proposals at the time, Lengerke's proposal reassured

---

21   Lengerke, George de. May 1944. *Airopia Ltd: A Company for the Putting into Operation of the European Part of the De Lengerke Plan for the Development of Civil Aviation after the War*, AIR 2/2508, National Archives, Kew.

22   Lengerke, George de. 13 September 1944. Letter to the Under-Secretary of State, AIR 2/2508, National Archives, Kew.

readers that international security would be catered for – any future international air police force would be allowed the use of the Airport Corporation's facilities.[23]

In April 1944 Count de Lengerke contacted the Air Ministry, and so began the process of applying for an operating licence and for permission to purchase York aircraft. The Ministry was not however impressed by Lengerke's proposal and spent the rest of 1944 and 1945 ensuring that he was unable to obtain both. One internal Ministry memorandum summed up the Air Ministry's attitude. The proposal was 'absurd and impracticable ... . Obviously Airopia cannot realise its pretensions to secure a monopoly of intra-European international services without the consent of all the European Governments concerned. This is hardly likely to be forthcoming'.[24] Airopia had 'bitten off more than they can chew. We have no real reason to repress them; but it seems pointless to encourage them in schemes which are almost certainly doomed to failure'. Financially, the company was 'pretty rickety'.[25] The Air Ministry had seen clearly that Airopia would be stalled by the growing national sovereignty of the European powers in the post-war period.

The foreign office agreed with the Air Ministry's assessment, but also expressed disquiet at the possibility of the appearance of aerial routes outside of the control of the British government. One official noted:

> the Company's plans are ambitious – to the point of fantasy. I cannot imagine how they expect to obtain the necessary equipment or the capital to acquire such an enormous fleet of aircraft. They do not seem to take into account either any international arrangements that may be concluded regarding the post-war organisation of civil aviation in Europe or the plans of old established national companies such as K.L.M. or Air France for post-war operation ... this high sounding scheme will die sooner or later of inanition.[26]

Another official hinted at a clash with Britain's foreign policy interests: 'we should not at all like to see an uncontrolled line running from Switzerland to Portugal now.'[27]

The Air Ministry delayed responding to Lengerke's request until the publication of the British government's White paper on the future of British aviation. This was published in December 1944, and essentially reserved all existing international

---

23   Airopia Limited. Memo. May 1944. *Compangne Europeene de Navigation Aerienne*, AIR 2/2508, National Archives, Kew.

24   AUS (C.A.). Memorandum. 11 September 1944. AIR 2/2508, National Archives, Kew.

25   Burkett, W.W. Memorandum to Air Chief Marshal W.R. Freeman. 13 September 1944. AIR 2/2508, National Archives, Kew.

26   Cheetham, N.J.A. Memorandum to J.H. Riddoch. 4 August 1944. AIR 2/2508, National Archives, Kew.

27   Hayter, W.G. Memorandum to W.P. Hildred. 2 December 1944. FO 371/42649, National Archives, Kew.

routes ploughed by British airlines for three government-owned British airlines.[28] This policy was disastrous for Airopia, and Airopia Director Leonard Plugge railed against it in Parliament in January 1945.[29] Lengerke was now forced to take his proposals in a radical new direction: he announced that his airline was now to fly only at night. This constituted flying on new routes, he argued, which consequently fell outside of the remit of the White Paper. For this he designed a special 'Sleeper of the Air', which was to carry passengers in compartments with beds, and took out a patent application.[30] In July 1945 he presented his *Service des Avions Lits et des Grands Express Aériens Européens* to the Air Ministry. This proposal jettisoned his earlier global vision and instead focused only on Europe, where Airopia was presented as an international European airline specialising in flying at night. Lengerke claimed that this would be popular with passengers wishing to avoid the expense of overnight stays in hotels during travel, and would allow for greater freight and mail carrying capacity.[31]

The Air Ministry, however, still remained opposed to Airopia. Lengerke consequently shifted his focus to the continent. His proposals were generally being positively received in press sympathetic to pan-Europeanist causes – 'Voilà donc un bien vaste et grand projet' exclaimed the Swiss *Le Confédéré*.[32] Lengerke had already opened discussions with other governments, hoping their approval would help turn British opinion in his favour. In June 1944 he had met the Turkish Air Attaché in London, in November 1944 the Swiss Air Attaché, and in late 1944 he flew to Lisbon to meet the Portuguese air authorities. These attempts however were unsuccessful – probably because these authorities sensed that Airopia did not have British government backing.[33] He renewed his efforts in late 1945. In December he visited Switzerland with the aim of basing his company there. He suggested that Swiss Air, in return for a generous fee, operate the airline on Airopia's behalf. Although Credit Suisse agreed to lend him two-thirds of his required capital, permission to run an airline in Switzerland was denied by the Swiss aviation authorities (one official condemned the proposal as 'fantastic'), and so these plans too came to nothing.[34]

---

28    December 1945. *British Air Services,* Cmd. 6712.

29    HC Deb 26 January 1945 volume 407 columns 1199–1202.

30    Lengerke, George de. Letter to William Hildred. 26 July 1945. AIR 2/2508, National Archives, Kew.

31    Airopia.25 July 1945. *Service des Avions Lits et des Grands Express Aériens Européens*, AIR 2/2508, National Archives, Kew.

32    1944. Nouvelles de l'étranger. *Le Confédéré,* 20 March.

33    Wellington, R.A. Memorandum to the Air Ministry. 3 January 1945. AIR 2/2508, National Archives, Kew. Growdon. Memorandum to the Air Ministry, 25 November 1944. And the attached memorandum from Swiss Air Attaché C. Schlegel to Air Commodore F. Beaumont, n.d., AIR 2/2508, National Archives, Kew.

34    Air Attaché, British Legation in Berne to Assistant Chief of Air Staff (Intelligence), Air Ministry. 15 February 1946 and 20 February 1946 and 26 February 1946. AIR 2/2508,

Yet the company had some support in Britain. Although its vision was idealistic, its status as a private company attracted support. A.V. Roe, the manufacturer of the York transport aircraft, was keen to diversify from government contracts, and sell to private buyers such as Airopia. This is even though it was unable to meet its government delivery targets. The Managing Director warned in December 1944 that if he was not allowed to sell aircraft to companies such as Airopia, then the Americans would, and so eventually capture the European market.[35] The Air Ministry refused to give permission, arguing that the Government's requirement for aircraft, still not fully met, should first be fulfilled. A year later, and A.V. Roe was still trying to sell to Airopia, claiming that it would soon overcome labour shortages and meet and surpass government production targets.[36]

Minister of Supply John Wilmott (and also in charge of Aircraft Production) was also keen that Airopia get off the ground with British-built aeroplanes. Once again the argument was made in terms of British interests. Writing to the Minister for Civil Aviation, he argued: 'If the Swiss company do succeed in establishing services … I think it will be a valuable advertisement to the British aircraft industry and go some way toward meeting Bevin's desire to see British aircraft in operation on the Continent as soon as possible.'[37] He suggested that six Yorks be diverted from RAF Transport Command's allocation to Airopia[38] – a suggestion the Minister of Civil Aviation took particular exception to. The Minister pointed out that Yorks were required by the government, and if Transport Command could spare them, then they should be allocated to commercial routes to South Africa, India, and South America, or sold to Brazil, which had been requesting them for many months. Brazil was 'potentially a far better customer for aircraft than a doubtful Swiss airline could ever be'.[39]

Lengerke's proposal was produced at a time when the British government was not only reorganising Britain's nationalised airlines, but also considering the possibility of allowing shipping and railway companies to participate in the running of Britain's airline networks. Although in November 1945 the new Labour government decided against this option (it opted instead for complete government ownership and control), there was nevertheless considerable debate as to the merits of this option. Lengerke and Labourite internationalists, although contending

National Archives, Kew. See also attached Memorandum by Count de Lengerke dated 14 February 1946.

35   Dobson, A.H. Letter to Stafford Cripps. 2 December 1944. AIR 2/2508, National Archives, Kew.

36   Roe, A.V. Letter to William P. Hildred. 14 December 1945. AIR 2/2508, National Archives, Kew.

37   Wilmott, John. Memorandum to Lord Winster.28 January 1946, AIR 2/2508, National Archives, Kew.

38   Wilmott, John. Memorandum to William Wedgewood Benn. 4 January 1946. AIR 2/2508, National Archives, Kew.

39   Lord Winster. Letter to John Wilmott. 16 January 1946. AIR 2/2508, National Archives, Kew.

primarily international European, and not British, aviation, nevertheless found themselves on different sides of the British debate. The 'Wings of Peace' and Airopia visions consequently clashed during Parliamentary debates on the future of British airlines, and even in the press. Lengerke in particular, was able to attack Labour internationalists' use of the Wagons-Lits metaphor in support of their vision of government-directed national and international aviation. The most public exchange began with a November 1945 parliamentary debate on the future of British civil aviation. Labour Whip Lord Strabolgi made a speech lamenting the Chicago outcome in the following way:

> We do aim at far more international cooperation in civil aviation. I hope I shall not be giving any of your Lordships nostalgic feelings when I say this, but I always think of the admirable service we had from the International Wagons-Lits in Europe in the old days. You could step from a cross-Channel steamer at Calais into your compartment and go to Moscow or Istanbul or wherever you wished over all frontiers, though all Customs stations on the international railway lines. That was possible in a far more difficult sphere than in the case of aviation.[40]

Lengerke's responded in *The Times* that Wagons-Lits was efficient because it was actually a private concern:

> The fact is that although the majority of Europe's railways were nationalized and State owned, it was found necessary to entrust a privately owned international company with the operation of long-distance international communications to ensure efficient and "admirable service." Surely what applied to the railways years ago applied equally well to the airways to-day?[41]

Strabolgi also responded in *The Times*, but, as the government had not publicly committed itself to any scheme for European aviation, chose to emphasise not the ownership of any internationalised aviation, but rather the ease with which it would operate internationally:

> I was not concerned as to whether they are or will be privately or publicly owned. The interesting feature of the International Sleeping Car Company was that its coaches passed freely all over Europe, and I see no reason why a properly organized international air service should not have the same facilities; nor do I suggest that the declared policy of the Government is in any way in conflict with this ideal ...[42]

---

40    HL Deb 01 November 1945 volume 137 column 646.
41    Lengerke. 1945. Letter to the Editor. Civil Aviation. *The Times*, 5 November.
42    Lord Strabolgi. 1945. Letter to the Editor. Civil Aviation. *The Times*, 8 November.

## 5. Conclusion

Although Wagons-Lits had been linked to visions of a centrally organised international aviation since the mid 1920s, it reached its greatest prominence during the war. This turn to the analogy of Wagons-Lits should thus be understood as an attempt by internationalists to inject new life into their long-existing proposals: a last ditch attempt to make use of the opportunities opened up through (as they perceived it) the peculiar malleability of world order that existed in the 1940s. Britain and America were aerially dominant over Europe and would remain so only for a few more years after the end of the war. That was the time, British internationalists believed, to mould aviation for the national and international interest. The analogy of Wagons-Lits was used to make their proposals more appealing to a wide audience. Railway companies, after all, were already being discussed as potential partners for aviation within Government. Wagons-Lits had transcended its interwar image of luxurious travel, and now appeared to stand for the future of international transport: rationally planned on an international scale, seamless across national boundaries, efficient, and an exemplar of apolitical international cooperation (that, in actuality, Wagons-Lits was none of these seemed to matter little to the internationalists). At the same time, the metaphor of Wagons-Lits provided a site for contestation between the two contending visions of aerial internationalism. Lengerke chose to emphasise private ownership as the source of Wagons-Lits' success. Supporters of an international organisation controlled by national governments, on the other hand, chose to simply emphasise the seemingly seamless nature and rationality of Wagons-Lits' operation.

These visions of internationalised aviation were fuelled by two further currents in British society and political culture. First there was the increasing assumption that integration in technical matters, and particularly transport, would bring war torn European countries closer together, and thus act as a force for peace. The wartime move towards nationalisation and national ownership, meanwhile, led many to envisage international aviation as an inter-government enterprise. This way of thinking about international aviation was in the ascendant until late 1945, when the Chicago conference on the future of international aviation put it to rest. The American government instead pushed for an aerial system dominated by nation-states.

Supporters of private enterprise put forward their own brand of internationalised aviation which came to the fore in 1944 and 1945. Set against the prevalent nationalist and nationalising mood, Airopia too failed however. As European nation-states reconstructed and reasserted themselves in the years following the war, British influence on aviation across Europe declined. As a consequence of this decline, proposals for internationally planned aviation disappeared from Britain. In their place, proposals for the integration of European aviation arose from within continental European countries. The most radical and prominent of these was the proposed merger between Lufthansa, Sabena, Air France and Alitalia – an 'Air Union' first proposed in 1957 but abandoned in 1968. Air Union, like

British proposals in 1944 and 1945, was driven by concerns over increased US competition, but in addition included a concern over the imminent widespread introduction of jet airliners (Dienel 2004).

The competing and contrasting visions of internationalised aviation described in this chapter have been located in differing political visions, principally private monopoly versus inter-governmental control. This chapter has rooted these internationalist visions deeply in the social, cultural and political currents of the day, whilst maintaining links to ideas emerging (or re-emerging) from the interwar years. By doing so, it has taken these visions one step further away from being thought of as simply utopian, and one step closer to the complexities of real-world political commitments. Looking beyond aviation, one important implication of this study is that describing technocratic, infrastructural or technological visions as being simply derivative of long-standing utopian impulses (e.g. Saint-Simonianism, which has sometimes been directly and unproblematically linked to technocratic projects, for example in Fischer 1990) masks the political ideologies and political commitments driving the visions. This chapter, thus, suggests that one important way to study such visions is to understand them as manifestations of contemporary political views. This would help us to better understand not only the visions and those who supported them, but also the underlying political viewpoints in new meaningful ways.

## References

Anastasiadou, Irene. 2007. Networks of Powers: Railway Visions in Inter-War Europe. *The Journal of Transport History* 28(2), 172–91.

Arnold-Forster, W. 1931. *The Disarmament Conference*. London: National Peace Council, 77–8.

Behrend, George. 1959. *The History of Wagons-Lits 1875–1955*. London: Modern Transport Publishing Co.

Behrend, George. 1962. *Grand European Expresses: The Story of Wagons-Lits*. London: George Allen & Unwin.

Bowles, F.G. and Major W.F. Vernon. 1946. *World Airways: What We Want*. London.

Bowles, Frank. 1944. M.P. Wants World Authority to Control Civil Aviation. *Evening Standard*, 16 March.

Brewin, Christopher. 1982. British Plans for International Operating Agencies for Civil Aviation, 1941–1945. *International History Review* 4(1), 91–110

Brossolette, Pierre. 1931. Sur un Projet de Réduction des Armements Aériens. *Notre Temps* 2(118), 29 November, 513–18.

Carr, E.H. 1942. *Conditions of Peace*. London: Macmillan and Co.

Devereux, David R. 1991. British Planning for Postwar Civil Aviation 1942–1945: A Study in Anglo-American Rivalry. *Twentieth Century British History* 2(1), 26–46.

Dienel, Hans-Liudger. 2004. *Seeking for Partners: Lufthansa's International Collaboration 1953–2000*. Paper to the 8th Annual Conference of the European Business History Association.

Douhet, Giulio and Dino Ferrari. 1942. *Command of the Air*. 2nd Edition. New York: Coward-McCann, 79.

Fischer, Frank, 1990. *Technocracy and the Politics of Expertise*. Newbury Park: Sage Publications.

Harper, Harry. 1929. *Twenty-Five Years of Flying: Impressions, Recollections, and Descriptions*. London: Hutchinson, 229.

Joyce, James Avery and Michael Young. 1945. *Chicago Commentary: The Truth about the International Air Conference*. London.

Kranakis, Eda. 2010. European Civil Aviation in an Era of Hegemonic Nationalism: Infrastructure, Air Mobility, and European Identity Formation, 1919–1933, in *Materializing Europe: Transnational Infrastructures and the Project of Europe*, edited by Alec Badenoch and Andreas Fickers. Houndmills, Basingstoke: Palgrave Macmillan.

Longmore, Arthur. 1946. *From Sea to Sky: 1910–1945*. London: Geoffrey Bles, 301–2.

Lyth, Peter J. 1995. The Changing Role of Government in British Civil Air Transport 1919–49, in *The Political Economy of Nationalism in Britain 1920–1950*, edited by Robert Millward and John Singleton. Cambridge: Cambridge University Press, 65–87.

Mance, H.O. 1942. Air Transport and the Future: Some International Problems. *The Bulletin of International News*, XIX(26), 1173–8.

Millward, Robert. 2005. *Private and Public Enterprise in Europe*. Cambridge: Cambridge University Press, 231–43.

Mitrany, David. 1943. *A Working Peace System: An Argument for the Functional Development of International Organization*. London: Royal Institute of International Affairs.

Parker, Percy Livingston. 1928. *Daily Mail Year Book for 1929*. London: Associated Newspapers, 22.

Schipper, Frank. 2008. *Driving Europe: Building Europe on Roads in the Twentieth Century*. Amsterdam: Eindhoven University of Technology.

The Labour Party. 1944. *Wings for Peace: Labour's Post-War Policy for Civil Flying*. London: The Labour Party.

Tree, Ronald et al. 1943. *Air Transport Policy: By Four Conservative Members of Parliament*. London.

Van Vleck, Jenifer L. 2007. The Logic of the Air: Aviation and the Globalism of the American Century, *New Global Studies* 1(1), 1–37.

Wallis, Keith. 2007. *And The World Listened: The Biography of Captain Leonard F. Plugge. A Pioneer of Commercial Radio*. Tiverton: Kelly Publications.

Wilson, Peter.1996. The New Europe Debate in Wartime Britain, in *Visions of European Unity*, edited by Philomena Murray and Paul Rich. Boulder, Colorado: Westview, 39–62.

World Airways Joint Committee. 1944. *Civil Aviation and World Unity: A Manifesto*. London.

Young, Michael Dunlop. 1944. *Civil Aviation*. London.

Young, Michael Dunlop. 1947. *Labour's Plan for Plenty*. London: The Labour Party, 125–9, 131–5.

Zaidi, Waqar. 2009. *Technology and the Reconstruction of International Relations: Liberal Internationalist Proposals for the Internationalisation of Aviation and the International Control of Atomic Energy in Britain, USA and France, 1920–1950*. Unpublished PhD thesis, London: Imperial College London.

Zaidi, Waqar. 2011. Aviation Will Either Destroy or Save Our Civilization: Proposals for the International Control of Aviation, 1920–1945. *Journal of Contemporary History* 46(1), 150–78.

Chapter 11

# Iron Silk Roads: Comparing Interwar and Post-war Transnational Asian Railway Projects

Irene Anastasiadou and Aristotle Tympas

## 1. Introduction

We now know that the building of – or the failure to build – transnational technological infrastructure has been of primary importance to the making of Europe (Schot and Scranton forthcoming). Some of the most influential interpretations of Europe did not emerge in the context of politics, but in the context of certain plans for the development of transnational transportation, energy, and communication networks (Tympas and Anastasiadou 2006, Van der Vleuten and Kaijser 2006, Van der Vleuten et al. 2007: 321–47, Lagendijk 2008, Schipper 2008, Anastasiadou 2012). In this chapter, we refer to studies that focus on the understudied, yet critical, role of institutions and initiatives active in what has been aptly called 'technocratic internationalism' (Schot and Lagendijk 2008: 196–217, Schipper and Schot 2011: 245–64, Schot et al. 2011: 265–89). The protagonists of this version of politics were not prime ministers and/or army generals. Rather, they were the managers and engineers of provisional or permanent committees, which were formed to pursue the interoperability of railways across national borders and against political and social barriers. Committees aiming at the advancement of transnational railway and other infrastructural routes were present across Europe from early on (Tissot 1998, Tissot 2003, Anastasiadou 2005, Schot et al. 2011).

Projects for transnational railway routes in Europe contained and advanced certain interpretations of Europe. In the past, we have studied the constant desire to link Greece to the transnational European network by way of a hybrid rail-ferry link across the Ionian Sea (Tympas and Anastasiadou 2006, Anastasiadou 2012). This connection would have allowed Greece to advance its own interpretation of and position within Europe, by gaining independence from its nineteenth and twentieth century Balkan neighbours. It was this link that was missing for the development of what many in Greece were promoting as the modern railway equivalent of the Ancient Egnatia Road. When finished, this Greek 'Iron Egnatia Road' would have been a key part of a transnational railway line linking Europe and Asia through a territory controlled by Greece.

In this chapter we turn our attention to another attempt at appropriating ancient history in the context of the pursuit of a transnational railway line. The emphasis here is on transnational railway versions of what is now described as the 'Iron Silk Road'. Such a railway connection between Europe and Asia would be the modern railway equivalent of the Ancient Silk Road. In our opinion, adding the Iron Silk Road story to that of the Iron Egnatia Road represents an initial step towards the extension and enrichment of the transnational history of technology. Some of its conceptual tools are applied in a broader context of comparative historical research. We expect that this step will be doubly beneficial. First, a transnational history of technology, developed in reference to the study of European tensions, will assist in understanding comparable tensions within Asia. Second, this understanding can generate feedback beneficial to the study of European history.

Let us bring the example of the concept of 'technocratic internationalism', which has been tried in the context of studies of the history of League of Nations (LoN) and United Nations (UN) initiatives and institutions for transnational infrastructure. This concept has been used to denote the important role of technical committees that sought to overcome political divisions between nation-states in order to advance transnational technological infrastructures. So far, the concept has been developed in reference to the European committees of the LoN and UN. Our chapter seeks to introduce the actions of the equivalent Asian committees of the UN. The focus is on versions of the Iron Silk Road, which were promoted by the United Nations Economic and Social Council for Asia and the Far East, ECAFE, founded in 1947 (Schaaf 1953: 463–81, ECAFE 1972, Lokanathan 1954). In 1974, the committee was renamed the Economic and Social Commission for Asia and the Pacific, ESCAP (Luard 1982, ESCAP 1987, ESCAP 1997a, 1997b, Luard 1994). We hope that our chapter will inspire scholars to consider studies comparing the actions of these committees with existing literature on the equivalent UN European committees, United Nations Economic Committees for Europe, UNECE.


## 2. 'From Peking, by railway': Asia in European Interwar Railway Plans

The long nineteenth century witnessed the expansion of railways on the European continent and in the United States (O'Brien 1983, Robbins 1998, Fremdling 2003: 209–21, Taylor and Neu 2003). Railway networks soon expanded to cover less developed parts of the world, such as Asia and Africa. This occurred within a context of colonisation and geopolitical competition between the Great European Powers (Mikhailoff 1900, Theroux 1975, Headrick 1981:129–41, 1988, Marks 1991, McMurray 2001, Kerr 2003: 287–326, Ahuja 2004, Kerr 2007, Patrikeeff and Shukman 2007, McMeekin 2010). From the beginning, railway construction in the European colonies followed the geopolitical agendas of the colonial powers (Betts 1985, Mikhailoff 1900, McMurray 2001, Kerr 2003, Divall 2010). At the beginning of the twentieth century, large-scale railway projects, such as the Trans-Siberian Railway (TSR) and the Baghdad Railway, had demonstrated the military importance of railways in the waging of war and the consolidation of colonial power. These projects prepared the ground for technocratic railway projects of the interwar years, which assumed a colonialist character. Let us consider the example of a project received by the newly established Transit Organisation of the League of Nations (Transit Organisation, 1921) in the interwar period.

The Transit Organisation of the LoN was established at the first intergovernmental conference on 'Freedom of Communication and Transit', held in Barcelona in 1921. Its aim was to promote openness in international communications and transit. To do so, it established a general convention on the International Regime of Railways in 1923 (League of Nations 1921, 1924). This constituted a general code, which defined the conditions that would facilitate international railway traffic. However, despite the general character of the convention and its universal aspirations, most of its articles were derived from European practice and were intended for application in Europe. European powers appeared hesitant to apply such principles to railway construction in their colonies (League of Nations 1921: 119). As a result, transnational railway integration in other parts of the world was limited. Several individuals, however, continued on to propose projects for the internationalisation of railways in Africa and Asia in the interwar years.

Carlo Enrico Barduzzi, a retired Italian diplomat, presented a memorandum for the reconstruction of the European railways to the directors of the Transit Organisation and the International Labour Office, ILO.[1] Barduzzi argued that the cause of the economic depression of the interwar years was the loss of Western European markets for industrial products in the USA and Japan, which had absorbed European products prior to World War I. He pleaded for the creation of railway arteries that would connect industrialised Europe with its Asian and African colonies. One such railway artery would cross Europe, connecting it to Asia and thereby constituting a Trans-Asiatic Railway.

This artery would connect London, Paris, Geneva, Milan, Rome, Bari, Valona, Thessalonica and Istanbul. Then, it would cross the Bosporus through a tunnel and continue to Haidarpacha (Turkey), Ankara, Erzurum, (Turkey), Tabriz (Persia), Tehran (Iran), Mashhad (Iran), Heart (Afghanistan), Quetta (Pakistan), New Delhi (India), Calcutta (India), Chittagong (Bangladesh), Rangoon (Burma), Martaban (Burma) and Saigon (Hanoi, today Vietnam). Thus, it would comprise a railway route of approximately 12,000km (Paneuropa ferroviaria archive LoN Box R 256: 5). This Trans-Asiatic Railway would consolidate Anglo-French interests in Eastern Asia. As such, it would constitute a competitive land route opposite the

---

1   Barduzzi, C.E., Mémorandum réserve pour un projet ferroviaire intereuropéen asiatique présenté à la commission des communications et de transit prés de la société des nations 1931 Archive International Labour Office; Barduzzi, C.E. Progetto di una paneuropa ferroviaria archive LoN Box R 256; Barduzzi, C.E. Projet pour une paneurope ferroviaire library of the United Nations Geneva.

Trans-Siberian Railway, which was controlled by the USSR. In addition, it would complement the unfinished Baghdad Railway route. The interests of Western Europe, Barduzzi argued, converged in Asiatic Turkey and Persia. For Barduzzi, Asia was suffocating under the Russian control of railway movement within the northern zone. Thus, action was necessary to stop Western Asia from becoming a field open for penetration by Bolshevik propaganda. Therefore, it was crucial to directly connect Persia, Afghanistan, India and Indochina with Western Europe (Barduzzi Paneuropa ferroviaria archive LoN Box R 256: 32).

The realisation of such a grandiose project would require great expenditures and significant construction works. While the railway would rely largely on existing infrastructure, new lines would have to be built in areas deprived of railway connections. The available Asian railways had incompatible technical characteristics. A section of the line in Southern Asia was of a broad gauge. Barduzzi emphasised the importance of adopting a uniform gauge along the entire route of the Trans-Asiatic Railway. The broad gauge, he argued, would be necessary for creating a more stable structure, which would be able to withstand the violent monsoons. Barduzzi proposed that the whole of the line, from Ankara to Saigon, be of 1.67 m gauge (Barduzzi, Paneuropa ferroviaria archive LoN Box R 256: 38).

China and Japan, Barduzzi argued, were countries with dense populations and long-existing civilisations. They would provide clientele for European industry and buy Europe's industrial goods in exchange for cotton, grain, tea, rice, tobacco and coffee (Barduzzi, Mémorandum réserve pour un projet ferroviaire intereuropéen asiatique: 9).

The director of the Transit Organisation dismissed the proposal. In the 1930s, Barduzzi presented a proposal for the construction of a Trans-Balkan and a Trans-Asiatic Railway to the Italian Minister of Public Works (Barduzzi 1930). This was also rejected.[2] The idea, however, of the Trans-Asiatic railway to link Europe to its Asian colonies would remain alive.

*2.1 Reaching Europe from Southeast Asia*

Efforts throughout the post-World War II years to promote the establishment of modern nation-states in Asia, which would be able to compete with industrialised Western European nation-states, were accompanied by extensive railway reconstruction and modernisation projects (Chinese railways opens Sinkiang line 1978: 945, Sambamoorthi 1978: 74–8, Puru 1983: 32–3, Yu Xin 1983: 528–9, Twenty-first century railroading will be profitable 1985: 491, Keijian 1986: 334–5, Headrick 1988: 53; Kerr 2007: 157–8). Amidst a politically volatile climate, initial attempts to promote regional integration occurred within the context of ECAFE, the United Nations Economic and Social Council for Asia and the Far East,

---

2   National archive of Rome, Segreteria particolare del duce carteggio riservato: N.X.R. sottofascicolo no. 2

established in 1947 (Schaaf 1953: 463–81, Lokanathan 1954, ECAFE 1972). In order to achieve its aim of economic development of the Southeast Asian region, ECAFE acknowledged the importance of improving transport and communication links within the region. It established an Inland Transport Committee, which was composed of three subcommittees: a railway subcommittee, a road subcommittee and a waterways subcommittee.

In 1967, after several years of preliminary activities the Inland Transport Committee of ECAFE recommended that a large railway network be created with the cooperation of all railway administrations in the region. This idea dated back to 1932, when Sir Cecil Clementi, Governor of the Straits Settlements, mentioned it in a speech at the opening of the new railway terminal in Singapore (UN keeps the Trans-Asian railway project alive 1979: 20). The objective of such a venture would be to eventually link countries in the region with those in Europe and Southeast Asia by a Trans-Asian Railway (TAR). The TAR would serve as a continuous railway artery from Singapore to Istanbul, thereby connecting the Asian countries with Europe and Africa by way of the most direct route (Husain 1971: 882). Disconnected countries would be connected via land transport infrastructure. Thus, a new avenue of communication to Europe, which would be able to outcompete maritime and coastal shipping, would be created (ESCAP 1972: 26). Proponents of the TAR argued that the connection of Southeast Asian countries with Europe would be beneficial for their own development (Feyeux 1970: 18–24). Furthermore, such a railway artery was expected to stimulate economic growth by increasing trade flows and promoting industrialisation of the countries through which the railway would pass. Finally, the early proponents of the project expected it to contribute to the social and cultural integration of the Southeast Asian region (UN keeps the Trans-Asian railway project alive 1979: 20–21, Chartier 2008: 6). The TAR would make use of existing tracks and installations. Additionally, missing links would be constructed and technical material and fixed installations standardised. The railway lines in the ECAFE countries, at the time, comprised a total of 170,000km. The planned railway artery would be only 14,000km long. Therefore, in the early 1970s, Husain (1971: 882) characterised the project as 'a rather modest undertaking'. Two-thousand (2,000) km of railway lines would need to be constructed. This would include 1,400km between Thailand and East Pakistan (today Bangladesh) and 600km between Iran and West Pakistan via Zarand (Iran), Kerman (Iran) and Zahedan (Iran). The modernisation and expansion of national networks was an important element of the TAR venture. Furthermore, train ferries would be an important aspect of the proposed artery (Husain 1971: 882).

The construction of the TAR presented a formidable challenge. This was due to the incompatibility of the networks, which had been inherited by the newly created nation-states from the colonial era. Three regions were identified: (1) of standard gauge (1,435 mm) in Iran, (2) of broad gauge (1,676 mm) in Ceylon (today Sri Lanka), India and Pakistan (Western Pakistan and Eastern Pakistan; today Bangladesh), and (3) of metric gauge in Burma (today Myanmar), Cambodia, Malaysia, Singapore, Thailand and Vietnam (Husain 1971: 882). The technical

**Figure 11.1    Image of the Van train ferry**

*Source:* Private photograph, Florian Müller.
Note: The train ferry crossing the Van lake in the eastern part of Turkey was one of the train ferries on the TAR route. It was placed into service in the 1970s, when the railway route that connects Turkey to Iran was led down. It has been operated since then connecting the cities of Tatvan and Van – and consequently the broader route of Turkey to Iran – through railways.

incompatibility extended to differences in breaking and coupling systems, variations in loading gauge and axle load standards. In order to solve the problem of railway traffic between standard and broad gauge tracks, ECAFE considered various technologies. To solve these issues, the large-scale use of freight traffic in containers became a defining component of the project (Husain 1971: 884). However, even this ad hoc technological solution to compatibility issues could not eliminate the obstacles of the regional differentiation of the Southeast Asian networks. In the region from Dhaka (today Bangladesh) to Parbatipur (today Bangladesh) and up to Singapore, within which 5,300km of lines were of metric gauge, the ISO container (8 ft. by 8 ft.) could not pass unless the loading gauge was favourable. This was, however, not the case. In regard to traffic and operation, the TAR would ultimately be comprised of two sections, which, though connected, would have differing potentials and conduct largely independent activities (Feyeux 1970: 22).

Difficulties also arose in the field of institutional arrangements. The settling of legal and administrative issues was necessary in order to minimise frontier formalities (ESCAP 1972: 26). Although there were a number of international conventions for the facilitation of international container transport, the ECAFE countries' status of participation in them was not encouraging. Australia, Brunei and the Khmer Republic (today Cambodia) were the only ECAFE countries which had subscribed to the Customs Conventions on Containers (1956). Proponents of the TAR expected its construction to motivate Southeast Asian nation-states to adhere to existing conventions and to draft trade agreements for the facilitation of freight traffic across frontiers (Feyeux 1970: 21).

The realisation of the TAR was also confronted with various economic challenges. In the 1960s and 1970s, the economic reality of the regions through which the railway artery would pass did not justify the investments necessary for the construction of the line. In the Southeast Asian region, a large percentage of the population worked in agriculture and cattle-raising. Local resources did not allow for the financing of the railway project and trade levels did not justify the corresponding investments (Feyeux 1970: 21).

As in the interwar years, visionaries were important proponents of the project. In 1980, the Railway Gazette refers to the death of Mr. S.S.M. Husain, chief of the ESCAP Transport and Communications Committee, as being a setback to hopes of establishing an Asian Railway Union in the 1980s:

> Syed Masood Husain was a visionary, sometimes espousing causes like the Trans Asian Railway from Istanbul to Singapore that stepped beyond the bounds of political and economic reality. But he had a firm belief – by no means misplaced – not only in the part which international rail links could play in breaking down geographical barriers that separate the countries of South-east Asia, but also in the benefits that could flow from closer ties between railway men in this vast region. (Setback to Asian railway union 1980: 179)

Political obstacles throughout the 1960s and 1970s hindered the realisation of the project (ESCAP 2007: 6, 49, Chartier 2008). Progress proved to be slow, especially during the Cold War years, when neighbouring countries were reluctant to participate (ESCAP 2007: 49). The closed frontier policy of the Burmese government, the political differences between Vietnam and its neighbours, and the chronic economic problems of the region rendered the construction of necessary links within the next decade unlikely (Levett 1979: 1). However, Levett (1979: 1) noted that 'political changes [were taking place] rapidly and it would be unwise to predict that those gaps [would remain] at the end of the century'. The region that would benefit most from the construction of a TAR was Southeast Asia 'and the project [was being kept] very much alive by ESCAP'.

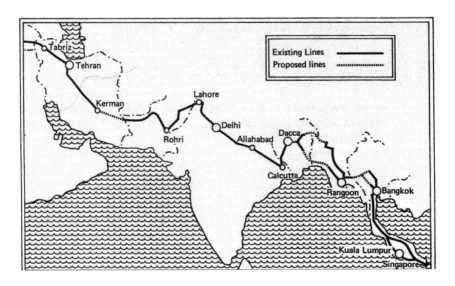

**Figure 11.2    ESCAP's Southern TAR**

*Source:* No Author 1979: 20.

*2.2 Proposals for a New Asian–European Land Bridge, 1980s*

While the realisation of the Southern TAR was confronted with serious obstacles, developments in the northern part of the Eurasian continent demonstrated the potential of railways in responding to the transport demands of the new era. Since the second half of the 1960s, the TSR established a Trans-Siberian Container Service (TSCS), which provided integrated container transport from Japan and Eastern Asia to Western Europe via rail, maritime and road infrastructure (ESCAP 1978: 45). The TSCS flourished in the 1970s and proved to be competitive opposite the maritime routes (Kolesnikov and Bodioul 1981: 27). Thus, it constituted the first land bridge between Southeast Asia and Europe. The concept of a land bridge was new and referred to the combined sea and land transport routes that used containers (Gopalan 1995: 18).

In 1981, V. Kolesnikov and V. Bodioul (1981: 23) presented the results of a study, in which they re-assessed the TAR venture as a land bridge between Southeast Asia and Europe. A hybrid route of rail and maritime transport would render the realisation of the TAR economically feasible and competitive opposite maritime routes. Until then, freight traffic between the Southeast Asian countries and their main trading partners – Western Europe, Japan and the USA – was carried almost exclusively overseas by fleets of the developed countries. As a result, the countries of the Southeast Asian region spent hundreds of millions of dollars on tonnage chartering (Kolesnikov and Bodioul 1981: 27). The construction of the TAR would enable the countries with railways to have their own means of transport to European markets. However, the construction of new railway routes would delay the realisation of the project. Furthermore, the transition from broad to metric gauge railways would render the eastern section of the network uncompetitive opposite maritime routes. Therefore, Kolesnikov and Bodioul (1981) thus argued in favour of the reassessment of the TAR as a land bridge based on the model of the TSR.

Container traffic, in the context of the TAR land bridge concept (Kolesnikov and Bodioul 1981), would use the TAR network from Western Europe to Iran, Pakistan and Bangladesh, and use maritime transport to reach Southeast Asia from the ports of Bandar Abbas (Iran), Karachi (Pakistan) and Calcutta (India). With combined railway and maritime delivery from Japan to Europe, as compared with transport through the Panama Canal, delivery times would be reduced by 3–7 days and transportation costs by 40 per cent. Still, the realisation of the TAR land bridge would require the construction of missing sections. Kolesnikov and Bodioul (1981: 25) argued that priority ought to be given to the Kerman-Zahedan line in Iran, 600km. Once the Kerman-Zahedan line was completed, a continuous railway would connect the Southeast Asian region and Europe. The railway would be part of the route from London to Calcutta, with a total length of over 11,000km.

In the first half of the 1980s, ESCAP acknowledged the importance of improving transport and communications within the Southeast Asian region and called upon governments of the region to declare a 'Transport and Communications Decade' (ESCAP 1985a: 1–2, ESCAP 1985b). The programme included work on the TAR that would be carried out on the principles of modernisation, the rehabilitation of existing railways, the improvement of traffic facilities at border crossing points where different systems met and new construction to close missing links (ESCAP 1985b: 20). However, until the 1990s, the improvement of existing infrastructure had priority over new construction. Still, as reported in the Railway Gazette, the Trans-Asian Railway from Istanbul to Singapore still lived in principle (Asian countries are coming together 1984: 163).

*2.3 Under Construction: A Pan-Asian Railway, 1990s*

The end of the Cold War and the new political order, which emerged in the final decade of the twentieth century, influenced the configuration of transnational railway routes for freight traffic. One of the most conspicuous developments was the decrease in the importance of the TSR in carrying containerised freight. The division of the transport system of the former USSR resulted in a decrease of freight traffic along the TSR, because of the emergence of new independent states and the economic and political problems of the Russian Federation (Kholopov 1997: 2). Additionally, the rise of China as an important economic power and the creation of new, independent landlocked countries in Central Asia provided the context within which proposals for the establishment of new land bridges – from Eastern Asia to Europe – were presented (ESCAP railway activities 1994: 15).

In September 1990, China's Beijing line, which connects Urumki (Western China) and Alashankou (on the western border of China), was linked to Kazakhstan Railways. The existing East–West lines of the national railways of Kazakhstan, Uzbekistan, Turkmenistan, Kyrgyzstan and Tajikistan were, therefore, connected with the railway tracks of the People's Republic of China. This was made possible by the collapse of the Soviet Union and the new independence of these five Central Asian countries. Instead of linking China to the Soviet Union, as was initially intended, 'the railway suddenly connected China to Central Asia' (Otsuka 2001: 42–9). Policymakers in China and Japan emphasised the importance of upgrading the line, so that it would constitute a transnational railway corridor connecting Eastern Asia to Western Europe, through Central Asia. The conceptualisation of the Silk Road was grounded on the fact that the new artery would follow a line parallel to the ancient Silk Road. The proponents of the project argued that the new artery – often referred to as the 'Modern Iron Silk Road' – would promote the economic development of China and offer the landlocked Central Asian countries access to new markets. The line would also contribute to economic and cultural exchange between Asia and the Pacific Nations, Europe and the Middle East (Shu 1997: 30–33). Despite the importance of the line the artery was far from achieving its maximum potential (Shu 1997: 31). To develop the railway into a truly international traffic corridor, Xu Shu notes (1997: 33), it was necessary to fully integrate the system so as to become more convenient, speedy and safe.

Beginning in 1992, ESCAP endorsed the TAR project. In its 48th session (April 1992), it noted that the development of transport and communications networks would constitute a major objective of Phase II (1992–1996) of the 'Transport and Communications Decade'. To meet the objectives in the area of land transport, ESCAP endorsed the Asian Land Transport Infrastructure Development (ALTID) project (ESCAP railway activities 1994: 16). Within the context of the ALTID project, the major principles of the TAR were revised to include routes in Northern and Central Asia. ESCAP carried out four corridor studies, which defined the routes of the TAR and the technical specifications to which the networks should adhere to (ESCAP 1996a. 1996b, 1996c, ESCAP 1999a, 1999b, ESCAP The Trans-Asian Railway no date). As a result, the TAR was expanded to cover the entire Asian region. China gained a prominent position in transit traffic with the newly defined Northern Railway Routes (NNR). In addition, the NRR and the 'North-South Corridor: Northern Europe to the Persian Gulf' would provide sea access to seven landlocked Asian countries. These included the newly established nation-states in Central Asia: Azerbaijan, Kazakhstan, Kyrgyzstan, Tajikistan, Turkmenistan, Uzbekistan and Mongolia.

Overall, these corridor studies identified 81,000km of routes of international importance, which served 28 member countries. ESCAP also promoted legal and administrative arrangements on an international level (ESCAP railway activities 1994: 16). An intergovernmental agreement on the TAR network was adopted by ESCAP on April 12, 2006. Later, the agreement was signed

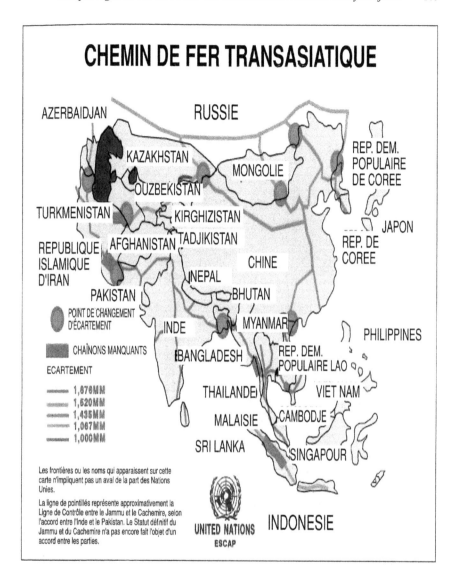

**Figure 11.3   A Pan-Asian railway**

*Source:* No Author 1994: 17.

by 22 member countries, of which 12 deposited their instrument of ratification, acceptance or approval with the Secretary-General of the United Nations at the UN headquarters in New York. The agreement entered into force on June 11, 2009. Under the terms of the agreement, a working group was established to consider its implementation and any proposed amendments (ESCAP Priority Investment Needs for the Development no date: 2).

In the last decade of the twentieth century and the first decade of the twenty-first century, the construction of Europe-Asia land transport links attracted the interest of the principal European transport bodies (Basiewicz et al. 1992: 33–7, ECMT 2005, 2006). The European countries established the Transport Corridor Europe-Caucasus-Asia programme to connect Central Asia with Europe along a continuous railway (Gorshkov and Bagaturia 2001: 50–55). Among the objectives of the project was the linking of this Eurasian transport corridor with European and global transport systems Gorshkov and Bagaturia 2001: 53).


## 3. Conclusion

This chapter constitutes an initial attempt to present the history of Asian railways from a transnational point-of-view. This view can be supported by taking advantage of a body of literature developed in reference to another continent: Europe. We find it appropriate to conclude with the presentation of a set of observations, rather than with a 'finite conclusion'. The first of these observations refers to the concept of 'technocratic internationalism' in the context of the UN committees that pursued transnational railways and other technological projects in the face of political division. According to the sketch of the history of versions of the TAR – as presented above – the role of these committees in Asia seems to have been limited. ECAFE, and later ESCAP, clearly endorsed some version of technocracy and internationalism when they promoted a project that contrasted sharply with the political divisions of post-war Asia. In this sense, these UN committees offered a vision of a transnational Asian railway, rather than a realisable plan for its implementation.

This initial impression is, of course, rather general and requires several qualifications. Additional study of the changing role of these committees amidst a fluid political environment is certainly necessary. To this end, it seems appropriate to note that ECAFE began as a committee for the far east of Europe (United Nations Economic and Social Council for Asia and the Far East). As such, it was overly shaped by the concluding phase of European colonialism. We have found ESCAP to function in a post-Cold War geopolitical environment containing strong Asian powers, some of which are already global superpowers. The closer we come to the present, the more we find versions of TAR that seek to reach Europe from Asia, rather than reaching Asia from Europe. China's success in implementing a version of the TAR offers a nice contrast to the failures to construct a similar railway during the Cold War period. The political context of the recent decades is different, with most Asian countries endorsing some version of a capitalist economy. Strong nation-states and minimal political convergence may be prerequisites for the success of technocratic internationalism.

From the technical perspective of the topic, which we have started to present in this chapter, it may be useful to note that the Southern TAR remained a project to link national networks. Jane Summerton has proposed a differentiation between three types of transnational infrastructures: territorial coupling of autonomous systems, full system integration and long-term alliances and linkages between systems (Summerton 1999: 109–10). The Southern TAR was never proposed as something more ambitious than the territorial coupling of autonomous systems.

While never actually built, the South Asian version of the TAR served an ideological force. It would be premature, at this point, to attempt any measurement of its impact or position in wider contexts (post-war colonialism, Cold War, post-Cold War). Studies assessing what the TAR meant, as a transnational project, in regard to the interpretation of Asia on a regional level, would be useful. This would be especially relevant considering the context of nation-states, which competed against one another by trying to advance competing interpretations of Asia and of Asia's relationship with Europe.

Railway historians have warned that the political visions embedded in, for example, colonialist and imperialist schemes of railway development were never realised. In Asia, similar to everywhere else, local cultures were influential in interpreting and modifying such schemes (Kerr 2007, McMurray 2001). The study of this influence, in regard to a transnational Asian railway project, would be very useful.

**References**

Ahuja, R. 2004. 'The bridge-builders': some notes on railways, pilgrimage and the British 'civilizing mission' in colonial India, in *Colonialism as Civilizing Mission; Cultural Ideology in British India,* edited by Harald Fischer-Tine and Michael Mann. London: Wimbledon Publishing Company, 95–116.

Anastasiadou, I. 2005. *International Railway Organization in 19th and 20th Century Europe* [Online: Transnational Infrastructures and the Rise of Europe]. Available at http://cms.tm.tue.nl/tie/files/pdf/WD.9.Anastasiadou.pdf.

Anastasiadou, I. 2012. *Constructing Iron Europe Transnationalism and Railways in the Interbellum.* Amsterdam: Amsterdam University Press.

Barduzzi, C.E. 1930. *La transbalcanica transasiatica: Roma, Bari, Durazzo, Stamboul, Haidar Pacha, Erzerum, Teheran, Kandahar, Lahore, Balkans-railway Bosphorus' tunnel and Transasiatic railway syndicates.* Faenza.

Basiewicz, T., Golaszawski, A., and Kopcinski, E. 1992. A European/Asian transport network concept. *Rail International* 5, 33–7.

Betts, R.F. 1985. *Uncertain Dimensions: Western Overseas Empires in the Twentieth Century.* Minneapolis: University of Minnesota Press.

Chartier, P. 2008. The Trans-Asian railway. *Transport and Communications Bulletin for Asia and the Far East,* 77, 1–24.

Chen, P. 1983. Matching growth to efficiency; Chinese railways. *Railway Gazette International*, 32–3.

Divall, C. 2010. *Railway Imperialisms, Railway Nationalisms*. Paper to the Travelling Public, Perceptions and Representations of Indian Railways: India.

ECAFE 1972. *ECAFE: 25 Years*. New York: United Nations.

ECMT 2005. *Globalisation: Europe-Asia Links; Synthesis Report and Political Decision Required*. Paris: ECMT.

ECMT 2006. *Transport Links between Europe and Asia*. Paris: ECMT.

ESCAP no date. *Development of the Trans-Asian Railway; Trans-Asian Railway in the North-South Corridor; Northern Europe to the Persian Gulf*. United Nations.

ESCAP no date. *Priority Investment Needs for the Development of the Trans-Asian Railway Network*. United Nations.

ESCAP no date. *The Trans-Asian Railway; Trans-Asian Railway in the North-South Corridor; Northern Europe to the Persian Gulf*. United Nations.

ESCAP 1972. The Trans-Asian railway network. *Transport and Communications Bulletin for Asia and the Far East*, 48, 25–7.

ESCAP 1978. Containerized haulage between Europe and Japan along the Trans-Siberian route. *Transport and Communications Bulletin for Asia and the Far East*, 52, 45–7.

ESCAP 1985a. Transport and communications decade for Asia and the Far East, 1985–1994. *Transport and Communications Bulletin for Asia and the Far East*, 56, 1–12.

ESCAP 1985b. *Launching of the Transport and Communications Decade for Asia and the Pacific, 1985–1994, Special Issue: Transport and Communications Bulletin for Asia and the Pacific*, 57, Bangkok: ESCAP.

ESCAP 1987. *ESCAP 1947–1987: Regional Cooperation for Development*. Bangkok: ESCAP.

ESCAP 1996a. *Trans-Asian Railway Route Requirements; Development of the Trans-Asian Railway in the Indochina and ASEAN Sub-region: Executive Summary*. New York: United Nations.

ESCAP 1996b. *Trans-Asian Railway Route Requirements; Development of the Trans-Asian Railway in the Indochina and ASEAN Sub-region: The Trans-Asian Railway in Indonesia, Malaysia, Singapore and Thailand (South)*. New York: United Nations.

ESCAP 1996c. *Trans-Asian Railway Route Requirements; Development of the Trans-Asian Railway in the Indochina and Asian Sub-Region; the Trans-Asian Railway in Cambodia, Southern China, the Lao People's Democratic Republic, Myanmar, Thailand (North and East) and Viet Nam*. New York: United Nations.

ESCAP 1997a. *ESCAP Today: Over Four Decades of Cooperation in Developing and Modernising the Asia-Pacific Region*. Bangkok: ESCAP.

ESCAP 1997b. *ESCAP, 1947–1997: 50 Years of Achievement*. Bangkok: ESCAP.

ESCAP 1999a. *Development of the Trans-Asian Railway; Trans-Asian Railway in the Southern Corridor of Asia-Europe Routes*. New York: United Nations.

ESCAP 1999b. *Development of Asia-Europe Rail Container Transport through Block-Trains; Northern Corridor of the Trans-Asian Railway*. New York: United Nations.

ESCAP 2003. *The Restructuring of Railways*. New York: United Nations.

ESCAP 2007. *The First Parliament of Asia: Sixty Years of the Economic and Social Commission for Asia and the Pacific 1947–2007*. Bangkok: United Nations.

Feyeux, M.M. 1970. The economic and technical aspects of the Trans-Asian railway network. *Transport and Communications Bulletin for Asia and the Far East*, 45, 18–24.

Fremdling, R. 2003. European railways 1825–2001: An overview. *Jahrbuch für Wirtschafts- Geschichte* 1, 209–21.

Gopalan, N.K. 1995. Modern transport systems in the globalized economy. *Transport and Communications Bulletin for Asia and the Far East,* 65, 15–26.

Gorshkov, T. and Bagaturia, G. 2001. TRACECA- restoration of silk route. *Japan Railway and Transport Review*, 28, 50–55.

Headrick, D.R. 1981. *The Tools of Empire: Technology and European Imperialism in the Nineteenth Century*. New York: Oxford University Press.

Headrick, D.R. 1988. *The Tentacles of Progress: Technology Transfer in the Age of Imperialism, 1850–1940*. New York: Oxford University Press.

Husain, M.S. 1971. Projet d'un réseau de chemin de fer trans-asiatique. *Rail International*, 882–4.

Kasuga, K. 1997. Trans-Asian railway. *International Cooperation,* 12, 31–5.

Keijian, L. 1986. Increasing coach production to meet China's needs. *Railway Gazzette,* 334–5.

Kerr, I.J. 2003. Representation and representations of the railways of colonial and post colonial South Asia. *Modern Asian Studies,* 287–326.

Kerr, I.J. 2007. *Engines of Change: The Railroads that Made India*. Westport, Connecticut: Praeger.

Kholopov, K.V. 1997. Tariff competitiveness of Trans-Siberian railway in Asia-Europe container transport. *Transport and Communications Bulletin for Asia and the Far East*, 67, 1–27.

Kolesnikov, V. and Bodioul, V. 1981. Development of freight traffic between Europe and the ESCAP countries: efficiency of rail-cum-sea route in the context of the Trans Asian railway network project. *Transport and Communications Bulletin for Asia and the Far East*, 54, 23–34.

League of Nations 1921. *Barcelona Conference. Verbatim Records and Texts of the Recommendations Relative to the International Regime of Railways and of Recommendations Relative to Ports Placed under International Regime*. Geneva: League of Nations.

League of Nations 1924. *Records and Texts Relating to the Convention and Statute on the International Regime of Railways; Leaque of*

*Nations, Second General Conference on Communications and Transit, November 15th–December 9th 1923*. Geneva: League of Nations.

Lagendijk, V. 2008. *Electrifying Europe: the Power of Europe in the Construction of Electricity Networks*. Amsterdam: Aksant.

Levett, J. 1979. Harmful tensions. *International Railway Journal*, 1.

Lokanathan, P. 1954. *ECAFE: The Economic Parliament of Asia*. Madras: Madras Diocesan Press.

Luard, E. 1982. *History of the United Nations: The Years of Western Domination 1945–1955*. London: Palgrave Macmillan.

Luard, E. 1994. *The United Nations; How it Works and What it Does*. London: Palgrave Macmillan.

Marks, S.G. 1991. *Road to Power: The Trans-Siberian Railroad and the Colonization of Asian Russia, 1850–1917*. Ithaca, NY: Cornell University Press.

McMeekin, S. 2010. *The Berlin-Baghdad Express: The Ottoman Empire and Germany's Bid for World Power*. Cambridge, MA: Belknap Press of Harvard University Press.

McMurray, J.S. 2001. *Distant Ties: Germany, the Ottoman Empire, and the Construction of the Baghdad Railway*. Westport, CT: Praeger.

Mikhailoff, M. 1900. The Great Siberian railway. *The North American Review*, DXXII, 593–608.

No author 1978. Chinese railways opens Sinkiang line. *Railway Gazette International*, 945.

No author 1979 (August). Un keeps the Trans-Asian railway project alive. *International Railway and Transit Review,* 20–21.

No author 1980. Setback to the Asian railway union. *Railway Gazette International*, 179.

No author 1984. Asian countries are coming together. *Railway Gazette International* 140 (11), 163.

No author 1985. 21st century railroading will be profitable. *Railway Gazette International*, 491.

No author 1994 (January). ESCAP railway activities. *Rail International*, 13–21.

O'Brien, P.K. 1983. *Railways and the Economic Development of Western Europe, 1830–1914*. New York: St. Martin's Press.

Otsuka, S. 2001. Central Asia's rail network and the Eurasian land bridge. *Japan Railway and Transport Review*, 28, 42–9.

Patrikeeff, F. and Shukman H. 2007. *Railways and the Russo-Japanese War*. London; New York: Routledge.

Puru C. 1983. Matching growth to efficiency; Chinese railways. *Railway Gazette International*, 32–3.

Robbins, M. 1998. *The Railway Age*. New York: Manchester University Press.

Sambamoorthi, R.M. 1978. Southern railway of India. *Railway Gazette International,* 74–8.

Schaaf, H.C. 1953. The United Nations economic commission for Asia and the Far East. *International Organization* 7 (4), 463–81.

Schipper, F. 2008. *Driving Europe: Building Europe on Roads in the Twentieth Century*. Amsterdam: Aksant.

Schipper, F. and Schot, J. 2011. Infrastructural Europeanism, or the project of building Europe on infrastructures; an Introduction, *History and Technology*, 27 (3), 245–64.

Schot, J., Buiter, H. and Anastasiadou, I. 2011. The dynamics of transnational railway governance in Europe during the long nineteenth century. *History and Technology*, 27(3), 265–89.

Schot, J. and Lagendijk, V. 2008. Technocratic internationalism in the interwar years: Building Europe on motorways and electricity networks. *Journal of Modern European History*, 6 (2), 196–217.

Schot, J. and Scranton, P. forthcoming, *Making Europe*. Palgrave Macmillan.

Shu, X. 1997. The new Asia-Europe land bridge: Current situation and future prospects. *Japan Railway and Transport Review*, 30–33.

Summerton, J. 1999. Power plays; the politics of interlinking systems', in *The Governance of Large Technical Systems,* edited by O. Coutard. London: Routledge.

Taylor, G.R. and Neu, I.D. 2003. *The American Railroad Network, 1861–1890*. Urbana: University of Illinois Press.

Theroux, P. 1975. *The Great Railway Bazaar, by Train through Asia*. New York: Mariner Books.

Tissot, L. 1998. Naissance d'une Europe ferroviaire: la convention internationale de Berne (1890), *in Les entreprises et leurs réseaux: Hommes, capitaux, techniques et pouvoirs, XIXe-XXe siècles: mélangés en l'honneur de François Caron* edited by *M. Merger and D. Barjot*. Paris: Presses de l'université de Paris-Sorbonne, 283–95.

Tissot, L. 2003. The internationality of railways: an impossible achievement', in *Die Internationalität der Eisenbahn: 1850–1970*, edited by M. Burri, K.T. Elsasser and D. Gugerli. Zürich: Chronos, 259–72.

Tympas, A. and Anastasiadou. I. 2006. Constructing Balkan Europe: The modern Greek pursuit of an 'Iron Egnatia', in *Networking Europe: Transnational Infrastructures and the Shaping of Eu*rope, edited by E. van der Vleuten and A. Kaijser. Sagamore Beach: Science History Publications, 25–49.

Van der Vleuten, E. and Kaijser, A. 2006. *Networking Europe: Transnational Infrastructures and the Shaping of Europe, 1850–2000*. Sagamore Beach, MA: Science History Publication.

Van der Vleuten, E., Anastasiadou, I., Lagendijk V. and Schipper, F. 2007. Europe's system builders: the contested shaping of transnational road, electricity and rail networks. *Contemporary European History*, 16 (03), 321–47.

Yu-Xin, Z. 1983. Unit coal trains feed China's economic development. *Railway Gazette International*, 528–9.

Chapter 12

# Spatial and Social Effects of Infrastructural Integration in the Case of the Polish Borders

Tomasz Komornicki

## 1. Introduction

The boundaries of contemporary Poland constitute a convenient subject of study of the mutual relations between the existence of political formal-legal borders, the development of infrastructure and the intensity of international (i.e. trans-border) socio-economic interactions. The subject is convenient as a result of the historical conditions and spatial instability of political boundaries. These factors were characteristic throughout the twentieth century, especially following the development of basic transport infrastructure. In the 1930s, R. Hartshorne, inspired by the changes that had occurred on the political map of Europe after World War I, proposed the classification of boundaries as either antecedent (i.e. preceding spatial development) or subsequent (i.e. secondary to such forms, see Hartshorne, 1936) boundaries. In the former case, trans-border connections developed more slowly than internal interactions within the neighbouring countries; the development of internal core areas dominated in autarchic economies. In the latter case, the already-established networks were cut across (e.g. in post-war Germany or the British Indies). During the interwar period, the newly-established and, therefore, subsequent Polish-German border in Silesia was the subject of a detailed study conducted by Hartshorne.

The purpose of this chapter is to define the interrelations between transport infrastructure, political boundaries and socioeconomic interactions. The assessment of these interrelations is demonstrated using the example of contemporary Polish boundaries and, in more detail, of two segments of the boundary. These segments are the boundaries between Poland and Germany and between Poland and Ukraine. The report makes use of earlier studies, conducted by the author for the Institute of Geography and Spatial Organization of the Polish Academy of Sciences (IGSO PAS). Diverse source materials were referenced: from the Polish Border Guards (data on cross-border traffic), the Polish Central Statistical Office, GUS (data on inward tourism) and the Customs Department of the Ministry of Finance (data on foreign trade).

Section 2 presents the mutual relationships between the development of infrastructure, its functions and the permeability of boundaries. Section 3 presents characteristics of the functional transformations of the Polish boundaries after World War II. Section 4 presents infrastructural developments, which took place after 1989, in the context of an intensive increase in traffic (Section 5) across all Polish boundaries. In this context, a short consideration of the current spatial distribution of Polish-German and Polish-Ukrainian interactions across the Polish territory is presented. Two international measures have been applied: bilateral trade exchange and inward tourism flow.

## 2. Infrastructure and Functions of Boundaries

The broadly conceived trans-border links are strongly associated with transport networks. These networks are platforms for passenger and commodity flows. Hence, international connections both depend upon and influence (i.e. strong relations promote new transport-related undertakings) transport systems. Some scholars consider transport routes to be interesting and representative in the spatial sciences, as they reflect forms of economic and social life (Clozier, 1959). E.L. Ullman proposed that transport be used as a measuring stick for interregional links. He stated that 'in order to establish the connections between individual areas and the essence of spatial exchange it is necessary to find a concrete manner of measuring and mapping the flow of cargo and persons, taking into account the magnitude and the speed of transport, as well as its origins and destinations' (Ullman, 1959).

The study of boundaries constitutes one of the fundamental subjects in political geography, history and, today, in economics and economic geography. Traditionally, geographers, following in the footsteps of historians, were interested in the course of political boundaries, their delineation and the reasons behind their concrete courses (see, e.g., Hartshorne, 1936; Boggs, 1940; Minghi, 1963; Prescott, 1967). Therefore, the key issue in the past was that of 'natural boundaries'. The analysis of border infrastructure was secondary at that time. It often constituted an element in the assessment of the correctness of boundary demarcation.

Beginning in the second half of the twentieth century, boundaries started to be perceived as spatial barriers. According to Kolosov (2005), the functional approach to the study of boundaries emerged in the early 1950s. At this time, the question of the degree of formalisation of these barriers was posed and the solely negative influence of a boundary on the growth base of adjacent areas questioned. W. Christaller (1963) maintained that central places, situated close to a stable boundary, could potentially profit from corresponding developments (e.g. of trade, warehousing, etc.). Sanguin (1983) distinguished three kinds of border-adjacent areas:

a. gaps (i.e. undeveloped areas);
b. cloudy areas (i.e. dispersed settlements with limited connections); and
c. border regions (i.e. areas with strong trans-border ties).

The course of borders ceased to be the main subject of study. This was replaced by the functions of borders. Boundaries were thereafter perceived as being evolving, 'mobile frontiers'. This implied the temporary character of boundary functions in regard to flows. Forecasts of the gradual reduction of flows amidst advancing spatial integration emerged (Corvers, Giaoutzi, 1998). The study of boundaries extended to different disciplines and the analytical scope broadened to encompass border-adjacent and trans-border areas (see J.Z. Garcia, 2003).

According to the available literature, the character of a border is determined by its degree of formalisation. The character of a spatial barrier can be described using indicators of permeability (Rykiel 1990). A boundary is understood as being a spatial barrier that influences the magnitude of the flow of persons and goods. This might be forecast on the basis of gravity models, in which the sole factor limiting interaction is the friction of distance. A jump-like break in the regression line – depicting mutual economic and social ties – is more pronounced with political boundaries, as these are stronger formal-legal barriers (Love, Moryadas, 1975). The weakening of this effect is significantly delayed. A jump-like decrease in intensity of spatial interactions can also be observed across borders that have been highly permeable for a longer period of time [e.g. between the Benelux countries (Rietveld, 2001) and between Germany and France (Helbe, 2007)]. The appearance of new boundaries leads to a jump-like increase of transaction costs. This is not only related to customs and associated fees, but also to the time of transport (Komornicki, 1999; Megoran et al., 2005).

In the past, the emergence of new political boundaries was an effect of geopolitical conflict, which frequently included military operations. However, the degrees of formalisation and, therefore, permeability of the emerging boundaries varied. Many of the boundaries that emerged after World War I, as a result of the Treaty of Versailles, remained formalised and permeable to a relatively low degree. This applied mainly to boundaries of the countries founded after the disintegration of the Austro-Hungarian Empire. The political divisions after 1945 resulted in the increased formalisation of boundaries along the Iron Curtain, but also between those countries of Central Europe that found themselves to be within the Soviet sphere of influence. The degree of formalisation impacted the situation of transport infrastructure. The weakening of economic and social interactions resulted in a decreased demand for transport and, therefore, for transport infrastructure. Some of the existing routes were no longer needed, others liquidated or not reconstructed for military reasons (Lijewski, 1994). When the interactions between the separated countries started to develop once more (i.e. beginning in the late 1950s in Central Europe), the demand for trans-border infrastructure increased. At the same time, however, high formalisation of the boundaries, resulting from political conditions, has been maintained. This has resulted in a specific cross-border transport infrastructure (e.g. border crossing and customs facilities, etc.) (Komornicki, 2010).

After Poland joined the Schengen Area in December of 2007, the dynamics of cross-border traffic at crossing points on the Polish border changed distinctly. Along the boundaries with Germany, Czechia, Slovakia and Lithuania, full permeability has been realised in a technical sense. However, permeability along the eastern border (i.e. with Russia, Belarus and Ukraine) has been limited anew. The role of the boundary as a formal-legal barrier has been strengthened. The full opening of the intra-Union borders was de facto the end of utilisation of the existing, specific cross-border transport infrastructure (Komornicki, 2010). Thereafter, the potential to develop mutual socioeconomic relations among the neighbouring countries depended upon the actual cross-border transport infrastructure. In this context, this refers to roads and railways that cross the boundaries and the quality (i.e. standards) of these routes. Geographical distance, after mental and socioeconomic aspects, continues to be a central challenge for trans-border cooperation in post-socialist European states (Herrschel, 2011).

Furthermore, we can consider whether the establishment of international connections in Poland, with neighbouring countries in particular, coincides with the establishment of the main transport routes within the country. It has previously been concluded that trade-transport interdependence exists (e.g. between Poland and Ukraine and between Poland and Germany) (Rösner et al., 1998; Komornicki, 2003).

## 3. Transformations of the Functions of Polish Boundaries and Their Impact on the Development of Infrastructure – A Historical Sketch

In the period of industrialisation and infrastructural expansion during the nineteenth century, the territory of Poland was controlled by three different European powers: Prussia (present Polish-German border), Tsarist Russia (present Polish-Belarusian border and the northern part of the Polish-Ukrainian border) and the Austrian-Hungarian Empire (southern segment of the Polish-Ukrainian border, Galicia). Each of these powers was at a different level of economic development and implemented different policies with respect to its peripheral areas, also in regard to infrastructural development (Lijewski, 1986). This resulted in a more intensive development in Prussian territories and a less intensive development in Russian territories. These differences were, to a certain degree, balanced throughout the twentieth century. However, they continue to be present in border-adjacent areas, especially the eastern areas, today.

During the interwar period, the functions of the individual boundaries of Poland were strongly differentiated. Likewise, their degree permeability varied. Several boundaries were relatively permeable: with Czechoslovakia (agreement on minor border traffic and 1926 tourist convention), with Germany (agreement on minor border traffic) and segments of the boundaries with Latvia and Romania. The border with the Soviet Union was less permeable and performed military, economic and social functions. The boundary with Lithuania remained completely closed (Komornicki, 1999).

Of the seven outer boundaries of present-day Poland, four have a typically subsequent character, as defined by Hartshorne in 1933. These are the boundaries with Germany, Ukraine, Belarus and Russia. Until the end of World War II, none of these borderlines were functional within the political sphere of Europe. The political order established in Europe after World War II included the extension of the Soviet Union (e.g. western Ukraine, western Belarus and the Baltic states) and the formation of a belt of countries, Poland included, which were subordinate to the USSR. Poland gained territory in the West, but lost a much larger territory in the East (Figure 12.1). The border to Germany was delimited on the basis of purely political criteria. In the case of the borders to Belarus and Ukraine, the complementing criterion was ethnic structure. Neither of these types of criteria could avoid cutting across transport infrastructure. In some situations, the negative consequences could be limited. The most extreme case was that of the present Polish border to the Kaliningrad district of the Russian Federation. This boundary was intended to divide the former East Prussian territory between the Soviet Union and Poland. On both sides of the border, a complete population exchange occurred. Moreover, the border was determined as

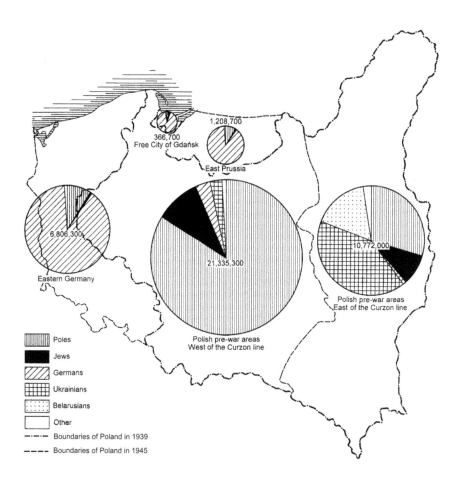

**Figure 12.1   Population structure, according to ethnicity, of areas between the eastern and western borders of Poland (established 1939 and 1946 respectively)**

*Source:* Eberhardt 2011.

a virtually straight line. Consequently, the scale at which the dense German-built transport network was cut through was the largest.

The Polish-German border is a late subsequent border, with reference to its course along elements of the natural environment. It was determined in 1945 by the Allied powers during the conferences in Yalta and Potsdam. Military activities of World War II had led to the destruction of all bridges across the Odra and Nysa Łużycka Rivers. Therefore, no transport infrastructure, which linked Poland and Germany, existed at the time. Reconstruction of the bridges progressed slowly. Even now, there are fewer such bridges than before World War II. Similar to the situation of the German border, population exchange occurred across the majority of the eastern Polish border after World War II. Many extensive forced-migration movements took place across the Polish-Ukrainian border. On the Ukrainian side, a nearly complete resettlement of the Polish population took place either during the war (e.g. deportations to the Soviet Union; Figure 12.1) or after 1945 (e.g. translocations to Polish areas that had not been incorporated into the Soviet Union). In the years following, the Ukrainian population living on the Polish side of the border was resettled. Initially, the resettlement moved in the direction of the Soviet Union. Later, as part of the so-called 'Action Vistula', resettlement moved towards the areas of Germany that had been taken over by Poland. Across the Polish-Belarusian border, post-war population exchange occurred on a smaller scale. A significant Polish minority remained on the Soviet side. On the Polish side, the remaining Eastern Orthodox population was much larger than the population that identified with the Belorussian nation. Across the Polish-Soviet border, delineated across East Prussia, population exchange occurred completely on both sides of the border (Eberhardt 2011).

Migration movements and the appearance of a tight border led to the breaking of functional social links and to serious economic limitations. This resulted in the hampering of the development of some towns located near the border. Grodno and Brest' in Belarus, Lviv in Ukraine and Przemyśl in Poland are all examples of towns that were separated from their catchment areas. Furthermore, some elements of transport infrastructure lost their significance. These were either closed down or limited to performing local functions. Most significantly, this concerned railway infrastructure of meridional orientation. The railway line Chełm–Brest–Białystok was cut across twice by the state boundary. Within the Polish-Russian borderland, much of the former German railway infrastructure had already been dismantled as part of the military operations. The materials which were 'extracted' were used for military purposes. Only some of these segments were later reconstructed.

After 1945, the military purpose of the Polish boundaries decreased significantly, as all of the neighbouring countries were official allies of Poland. Furthermore, the 'border zone' was delimited along all of the boundaries, including the eastern border. Special regulations, which were highly burdensome for both the local population and for visitors, were in force in these areas. The economic function of all the boundaries, despite official economic cooperation, preserved their significance. Immediately after World War II and until approximately 1949, cross-border population movements continued. This was associated with the resettlement of Germans and Ukrainians, and the repatriation of Poles from the eastern lands. After 1949, the borders of Poland were almost entirely closed. Extremely rigorous passport issuance policy was conducted. Still, within this same period, trade with the new neighbours increased gradually. The slow liberalisation process of the border regime began after 1956. Passport issuance regulations, such as those concerning the border zone, were slightly relaxed. An agreement on local cross-border traffic was signed with Czechoslovakia. Cross-border traffic was slowly increasing. Already then, daily commuting to Czechoslovakia and the GDR was beginning to emerge. Passport-free traffic between the socialist countries was introduced, while strict currency exchange limitations were preserved. Getting a passport for travel to Western Europe was facilitated to a great extent. In fact, travel beyond the Iron Curtain became much easier in and from Poland than from the other socialist countries. In practice, documentation of the possession of 150 USD was sufficient. Meanwhile, the number of tourists visiting Poland increased. Primarily, these were citizens of the Federal Republic of Germany. In the 1970s, trade value grew rapidly as well. This included trade with Western Europe. These changes were reflected, in part, through the opening of new border crossing points. They were established along the western and southern boundaries of the country. The situation changed least along the eastern border with the Soviet Union.

The emergence of the 'Solidarity' movement and of supply problems in Poland, which resulted in mass shopping by Poles abroad, in 1980 brought about the gradual withdrawal of socialist countries from the agreements with Poland on passport-free travel. During the first half of 1981, it was already easier to travel from Poland to Western Europe than to its neighbouring countries. In December of 1981, the boundary situation underwent a jump-like return of formalisation with the introduction of martial law. There was a drastic drop in traffic and some border crossing points were closed. Since 1983, passport regulations have again become increasingly liberal.

In the period of Germany division after World War II, separate regulations influenced the travel of Polish citizens to the former German Democratic Republic, to West Germany and to West Berlin. Travel to the present-day eastern States of Germany underwent far-reaching liberalisation between 1972 and 1980. In 1972, the border was opened. Crossing of the border on the basis of a stamp in an ID document and, initially, unlimited exchange of currencies was possible. Traffic to and from the Federal Republic of Germany was subject to a visa regime. Still, traffic in this direction was significant and, in the period after 1983, could be referred to as 'mass traffic'. Getting a West German visa in Poland was not a problem. The passport-related regulations were comparable to those of other western countries. During the last years of the People's Republic of Poland (PRL), travel to West Germany was much easier than to the GDR. Conform to international agreements, visa-free traffic was guaranteed in regard to travel to West Berlin. The limitations were constituted by Polish passport regulations.

Throughout the existence of the People's Republic of Poland (PRL, 1945–1989), the entire boundary with the Soviet Union was characterised by a very low degree of permeability. It constituted a kind of 'Second Iron Curtain', which separated the socialist countries of Central Europe from their 'Big Brother' (Komornicki, Miszczuk, 2010). In 1939, 63 railway and road routes crossed the line of the later boundary with the USSR. By the early 1980s, only two road and three railway crossing points linked the PRL and the Soviet Union across this border of 1,310km. There were a few other crossing points at which cargo and military were transported across the border. Along the Polish-Ukrainian border, all railway lines were closed for passenger traffic, with the exception of the Rzeszów-Lviv line. Until the late 1980s, only one crossing point for road traffic, at Medyka, functioned along this border. Officially, Polish-Soviet agreements on visa-free traffic were in force. In practice, however, one had to show a certified invitation or be a participant of an officially managed excursion (i.e. show a voucher issued by one of a few licensed state travel offices) in order to cross the border.

The ultimate liberalisation of the border regime in Poland occurred during 1989 and 1991. It resulted, finally, in the signing of agreements on visa-free traffic with all Western European countries and on local cross-border traffic with Czechia, Slovakia and Germany. Poland's share of international trade, together with a shift in the main directions of Polish foreign trade, increased tremendously. A dramatic increase of cross-border traffic across all borders was noted. This was partly caused by the mass phenomenon of travel to Poland to do shopping in the neighbouring countries. Initially, visitors from the former USSR had the intention of selling local products more cheaply. The increase in cross-border traffic was also due to the developments in transit and in foreign tourism among Poles. The new independent states, having emerged from the disintegration of the Soviet Union in 1991, became the successors to the agreement on visa-free traffic. After 1989, the interpretation of the Polish-Soviet agreement on visa-free travel changed. The voucher, which served as the confirmation for having paid for necessary tourist services (e.g. accommodation), could be issued even by small private companies. Consequently, such documents became generally available. Thereafter, the boundary became fully permeable for citizens of Poland and of all other descendant countries of the late Soviet Union.

In the years following, Poland signed new agreements with Lithuania and Ukraine, which were based on principles similar to those of the agreements with the countries of Western Europe. In the case of Russia and Belarus, old agreements persisted as valid until October of 2003. It was then that Poland, bound by obligations towards the European Union, renounced the old agreement(s) with Russia and Belarus, as well as the new one with Ukraine. A visa regime for the citizens of these countries was introduced. In a countermove, Russia and Belarus introduced visa regulations for Polish visitors. Ukraine, however, refrained from introducing visa regulations. In December of 2007, Poland and Lithuania joined the Schengen Area. The borders to Russia, Belarus and Ukraine became the boundary of this zone.

## 4. Development of Infrastructure after 1989

During the entire period of pre-accession transformation (1989–2004), specific infrastructure was established, primarily along the Polish boundaries. New border crossing points were opened along the previously closed routes. The construction of border crossing facilities was rarely accompanied by realistic transport-oriented investment projects. The modernisation of roads often encompassed only the road segment adjacent to the border. New roads crossing the border were hardly constructed. Germany was an exception. In the 1990s, a sound modernisation of the motorway network within the former German Democratic Republic was carried out. This modernisation included all four routes leading to the Polish border: Berlin-Szczecin, Berlin-Frankfurt, Cottbus-Forst and Dresden-Görlitz. Many of the roads in Poland, which led towards the boundaries, were rapidly devastated due to an increase and intensification of heavy loads traffic.

In regard to railway infrastructure, the 1990s witnessed a fast reactivation of the majority of the suspended connections. The trans-border passenger connections were put into motion and were, initially, quite popular. This tendency reversed quickly. In local trans-border relations, railways lost in the competition against road transport, which included rapidly developing international coach transport. Connections across the eastern border were being closed, due in part to expectations of customs inspection difficulties. The mass of goods hidden within elements and compartments of railroad equipment was subject to an excise tax. The single important railway investment of cross-border dimensions was the gradual modernisation of the Warsaw–Poznan–Berlin railway line. The conditions of the approaching accession to the European Union included investments into specific infrastructure. This was justified along the eastern border. In the cases of the German, Czech and Slovak borders, the level of investment in checkpoints was probably too high, as some facilities stopped functioning only a few years later (i.e. when Poland joined the Schengen Area). A change in the described situation occurred only with the appearance of external forms for financing transport projects, such as European Union funds. The very first projects, co-financed by the EU, appeared along the German border in the period preceding Poland's EU membership (e.g. ring road around border town of Guben-Gubin and bridge across Nysa River; co-financed using PHARE funds). Larger road and railway projects were supported by ISPA, the pre-accession programme. Later, this role was taken over by the Cohesion Fund and, since 2007, by the Operational Programme Infrastructure and Environment. At the same time, the western segment of the Warsaw-Berlin motorway was constructed on the basis of a concessionary system (BOT: Build – Operate – Transfer).

Despite the physico-geographical barrier (i.e. Odra and Nysa Łużycka Rivers), the present-day Polish-German border is well-equipped with road and railway infrastructure. After 1989, over a period of some five years, border crossing points were established on nearly all roads crossing the border. The policies of authorities of the border-adjacent German States supported, in the 1990s, the development of many local border crossings. Poland tried to concentrate on priority projects, in view of the limited means for such undertakings. Local communities also supported the opening of new border crossings. These perceived, in their existence, an opportunity for the development of trade and, thereby, of the entire local economy. This view was shared by the authorities of the four Polish-German Euroregions (Ciok 1994). Altogether, six generally accessible road border crossing points were established between 1990 and 1997. On the eve of the accession of Poland to the European Union, the use of existing road infrastructure was very high (90 per cent). Hence, it can be concluded that the infrastructural barrier, which existed along the border to Germany, turned into a barrier of low capacity of the main transport routes (i.e. not the barrier of the low number of border crossings).

After 2004, two new motorways in Poland were connected to the border. Thereby, transport on roads of the highest standard is now possible from the German border – and therefore from the German motorway system – to Cracow via Wrocław (A4 motorway) and to Lodz via Poznań (A2). At the same time, local projects were being implemented (e.g. walking bridge in Görlitz-Zgorzelec). The development of border crossing points between Poland and Ukraine occurred during the 1990s. It was then that comparably large facilities were put to use between Warsaw and Kiev and Lublin (Dorohusk) and between Warsaw and Lviv (Hrebenne). Later, two smaller facilities were established. A larger facility in Korczowa started taking over the main transit traffic between Lviv–Cracow–Dresden. The Polish A4 motorway is now being constructed towards the crossing point in Korczowa. The scale of investment projects along other routes towards the borders, of both road and railway, is much smaller. Two new local crossing points are now under construction. Due to the persistence of a high degree of formalisation of the border, transport infrastructure itself is not the crucial element determining permeability and the development of bilateral interactions in this case. The waiting times for customs and border clearance are very long. This is not due to the number of border crossing points, but to inspection procedures. Clearance procedures are conducted separately by Polish and Ukrainian officers. The scope of inspection is partly a reaction to small-scale trafficking of alcoholic beverages and tobacco products. No walking traffic is allowed, except at the Medyka crossing point. Within Europe, this is a highly exceptional situation.

## 5. Border Traffic

The number of persons crossing the Polish-German border annually increased rapidly during the first half of the 1990s. This trend was associated primarily with the intensive development of peri-border trade. During the period of the highest borderland prosperity (1994–1999), traffic in both directions exceeded 130 million persons each year. This traffic was clearly dominated by the citizens of Germany (Komornicki 1996; Figure 12.2). After 1999, a breakdown in the magnitude of traffic occurred. This affected foreigners more so than it affected Poles. It was caused by the balancing of retail prices in Germany and Poland and by the appearance of large

Warsaw and Lublin to Lviv and Kiev; Komornicki, 2003). The eastern borderland has been a region of strong concentration of exports to the immediate neighbours since the early 1990s. This results largely from the lower quality requirements posed by Russian, Ukrainian and Belarusian markets. Many petty business people from eastern Poland were not capable of competing in markets of the European Union and, therefore, sold their products in the East. The significance of the Ukrainian market for the eastern provinces of Poland has been decreasing since 2003. Meanwhile, there has been a pronounced increase in exports to Ukraine from the remaining Polish regions (e.g. western and northern Poland). This has definite consequences for transport policy. The connections between the centres of eastern Poland and Western Europe no longer cater exclusively to transit needs. Rather, they also serve Poland's economic activities. The spatial distribution of Ukrainian tourism in Poland (Figure 12.5) indicates that the highest concentrations of bilateral interactions are located in the largest Polish centres (e.g. Warsaw, Cracow, Gdansk and Wrocław). The Podkarpackie and Lubelskie provinces display a high percentage of Ukrainians among foreigners staying overnight. There is also a higher absolute number and share of overnight stays of Ukrainians in areas along the German border. This can be interpreted as the taking advantage of last opportunities for cheaper accommodation when traveling.

## 7. Summary

The analysis conducted demonstrates that mutual relations between functions of the boundaries, trans-border socioeconomic interactions and transport infrastructure are diverse and frequently assume a feedback character (Figure 12.6). The scope of these relations is variable and depends upon historical, geopolitical and economic factors. The dynamics of the processes taking place are distinctly different along boundaries within the European Union (i.e. tendency towards decreasing formalisation and increasing of permeability) than along the outer boundaries. In the former case, the Polish-German border being an example, the following phases have been distinguished:

* *Phase I (until approximately 1990)* was characterised by strongly formalised boundaries of centrally planned economies. The political and, to some extent, economic factors were decisive. Infrastructure, having appeared earlier, was used in a selective manner. Along with the development of economic relations and during periods of increasingly liberal passport and visa policy, this infrastructure became insufficient. The boundary, being an important spatial barrier, limited the development of infrastructure. Economic factors stimulated this development, but their force was too weak to balance the political factors (Figure 12.6, Phase I);
* *Phase II (1990–2007)* was a period of systemic transformation. A jump-like reduction of the limiting geopolitical factor occurred. The boundary became

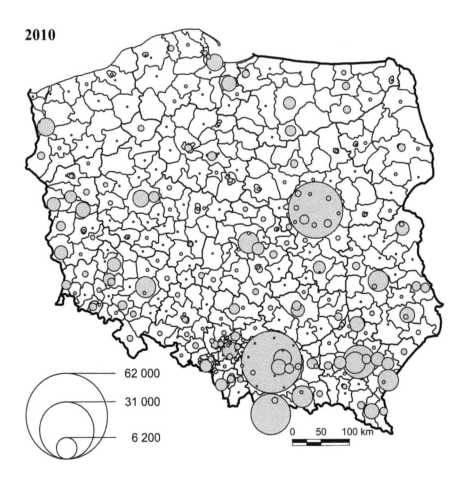

**Figure 12.5   Overnight stays of Ukrainian citizens in Poland**
*Source:* Elaboration (Komornicki) based on data from GUS.

increasingly permeable. Specific trans-border transport infrastructure developed. During definite periods, the development of economic relations, including transit, and of social relations resulted in a discrepancy between infrastructure and actual requirements. In terms of local relations, the improvement of the situation was relatively quick, due to simultaneous liberalisation of traffic and construction of border crossing points. The opening of the boundary was conducive to a general development of socioeconomic interactions. These, in turn, implied the need for the infrastructure on a European scale (Figure 12.6, Phase II). The gaps in modern infrastructure became a limiting factor for the integration processes;

- *Phase III (after 2007)* refers to the development of the inner boundary of the European Union and the Schengen Area. The accession to the European Union has brought about a further, rapid intensification of socioeconomic ties of trans-boundary character (Figure 12.6, Phase III). At the same time, the specific transport infrastructure has become useless. The availability of structural funds of the European Union has facilitated the development of the realistic trans-border road and railway infrastructure.

In summary, in the case of the Polish-German border, the functions and formalisation of the boundary led to developments of interactions and of transport infrastructure (negative and positive in the various phases, respectively) during both the initial and current phases of development. The consequence of infrastructural gaps as a hampering factor for the development of interactions was most distinct in the transformation phase.

In the case of the outer boundaries of the European Union, the process has taken a different course. The following phases have been identified:

- *Phase I* was characterised by a strongly formalised boundary, despite official political and economic cooperation. A clear separation of the economic functions (e.g. high intensity of trans-border transport) and social functions (e.g. very tight spatial barrier) occurred. The development of infrastructure, in the first case, was subordinate to economic factors. In the second case, this development was subordinate to geopolitics. Simultaneously, mass transport flows (i.e. economies based on heavy industries) limited the demand for border infrastructure (Figure 12.6, Phase I);
- *Phase IIA (1991–1997)* was a period of unevenly proceeding systemic transformation. The processes, initially, were similar to those along the western border of Poland. The geopolitical changes and the development of socioeconomic relations applied pressure on infrastructural development (e.g. border crossing facilities; Figure 12.6, Phase IIA);
- *Phase IIB (1997–2003)* continued to be a period of unevenly proceeding systemic transformation. Economic transformation occurred much more quickly in Poland. This led to price differences of consumer goods, which were not justified in market terms. Petty trade developed and exerted, with time, a limiting influence on the extension of infrastructure. Strong pressure for the establishment of new border crossing points took on a local character (Fifure 6, Phase IIB). Central authorities in both neighbouring countries started to become increasingly sceptical towards such initiatives (i.e. in view of the approaching EU membership of Poland). The first symptoms of renewed formalisation of the eastern border began appearing;
- *Phase III (after 2003)* refers to the development of the outer boundary of the European Union. The boundary has undergone formalisation in an abrupt manner once more. This does not entail the breaking of socioeconomic interactions, but rather their movement away from the boundary. In a local

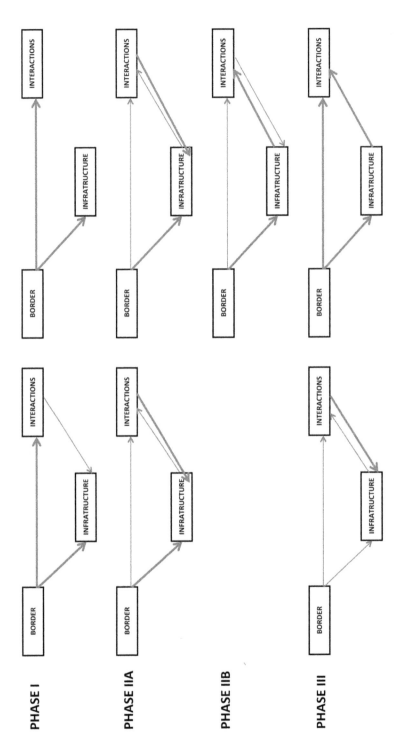

**Figure 12.6 Interrelations between functions of the boundaries, infrastructure and socioeconomic interactions**
*Source:* Elaboration (Komornicki).

sense, the boundary has become less permeable. At the same time, trade volume has increased and intensive flows of persons through the boundary have persisted. In part, this is due to the attractiveness of the Polish labour market for Ukrainians. The geopolitical factors are again influencing, in a limiting manner, the development of infrastructure. The influence exerted on bilateral interactions is primarily structural (Figure 12.6, Phase III). The demand for specific infrastructure has dropped, as this infrastructure does not resolve the issue of permeability. Simultaneously, demand for large projects has increased. However, such projects are not being carried out. The difficulties of the 2011 state budget have led to the elimination of routes, mainly, in eastern Poland from the plan for road construction until 2015.

Altogether, the processes occurring along the eastern Polish border seem to be more ambiguous. Geopolitical factors no longer determine the bilateral interactions, but continue to influence the demand for infrastructure. Transport infrastructure has limited, to a lesser degree over the entire period considered, the bilateral relations here than along the German border.

**References:**

Boggs S.W., 1940. *International Boundaries: A Study of Boundary Functions and Problems*, New York, Columbia University Press

Christaller W., 1963, *Ośrodki centralne w południowych Niemczech* (Central places of Southern Germany; Polish translation), PZLG, 1, IG PAN, pp. 1–72.

Ciok S., 2004, *Pogranicze polsko-niemieckie. Problemy współpracy transgranicznej* (Polish-German borderland. Problems of transborder cooperation; in Polish), Wydawnictwo Uniwersytetu Wrocławskiego, Wrocław.

Clozier R., 1959, *Geografia komunikacji* (Geography of transport; Polish translation), PZLG, 1, IG PAN, Warszawa, pp. 65–74

Eberhardt P., 2011, *Political migrations on Polish territories (1939–1950)*, Monografie 12, IGiPZ PAN, Warszawa.

Garcia J.Z., 2003, Directions in border research: an overview, *The social science journal* 40, pp. 523–33.

Hartshorne R., 1936: Suggestions on the terminology of political boundaries. In: *Annals of the Association of American Geographers*, 22, pp. 48–9.

Helbe M., 2007, *Border Effect Estimates for France and Germany Combining International Trade and International Transport Flows*, Kiel Institute, DOI:10.1007/s10290-007-0116-x.

Herrschel T., 2011, *Borders in Post-socialist Europe. Territory, Scale, Society,* Border Regions Series, Farnham, Ashgate.

Kolosov. V., 2005, Border studies: Changing perspectives and theoretical approaches, *Geopolitics* 10, pp. 606–32.

Komornicki T., 1996, *Trans-frontier traffic between Poland and Germany – a study of the situation at the frontier crossing points three years after the full opening of the border*, Arbeitsmaterial Nr.232 'Planerische und raumordnugsrechtliche Betachtung der grenzüberschreitenden Zusammenarbeit aus polnischer und deutscher Sicht, Akademie fur Raumforschung und Landesplanung, Hannover, pp. 144–55.

Komornicki T., 1999, *Granice Polski. Analiza zmian przenikalności w latach 1990–1996* (Boundaries of Poland. Analysis of changes in permeability in the years 1990–1996; in Polish), Geopolitical Studies vol. 5, IGiPZ PAN, Warszawa.

Komornicki T., 2010, *Transborder transport, w: A dictionary of transport analysis* (ed. K. Button, H. Vega, P. Nijkamp), Edward Elgar Publishing Limited, Cheltenham.

Komornicki T., Miszczuk A., 2010, Eastern Poland as the borderland of the European Union, *Questiones Geographicae* 29 (2), ed. P. Churski, Wydawnictwo Naukowe UAM, Poznań, pp. 55–70.

Lijewski T., 1986, *Geografia Transportu Polski* (Transport Geography of Poland; in Polish), PWE, Warszawa.

Lijewski T., 1994, *Infrastruktura komunikacyjna Polski wobec zmian politycznych i gospodarczych w Europie Środkowej i Wschodniej* (Transport infrastructure of Poland in view of the political and economic changes in Central and Eastern Europe; in Polish), Zeszyty IGiPZ PAN Nr 23, Warszawa.

Lowe J.C., Moryadas S., 1975, *The Geography of Movement*, Houghton Mifflin Company, Boston, pp. 333.

Megoran N., Raballand G., Bouyjou J., 2005, Performance, Representation and the Economics of Border Control in Uzbekistan, *Geopolitics* 10, pp. 712–40.

Minghi J.V., 1963, Boundary studies in political geography, in: *Annals of the Association of American Geographers*, 53, pp. 407–28.

Prescott J.R.V., 1967, *The Geography of Frontiers and Boundaries*, London.

Przestrzenne zróżnicowanie międzynarodowych powiązań społeczno-gospodarczych w Polsce (Spatial differentiation of international socio-economic connections in Poland; in Polish), *Prace Geograficzne*, 190, IGiPZ PAN, Warszawa, pp. 255.

Rössner T., Anisimowa G., Komornicki T., Miros K., Röttger A., 1998, Die Mitteleuropäische West-Ost-Achse Sachsen-Schlesien-Galizien, in: *Daten Fakten, Literatur Zur Geographie Europas* 5/1998, Institut fur Landerkunde, Leipzig, pp. 188.

Rietveld P., 1999, Obstacles to openness of border regions in Europe, typescript of a lecture, pp. 19.

Rykiel Z., 1990: *Koncepcje granic w badaniach geograficznych* (The concepts of boundaries in geographical studies; in Polish), in: Przegląd Geograficzny, 62, 263–73.

Sanguin A.L., 1983, *L'architecture spatiale des frontiers politiques: quelques reflections theoriques a propos de l'exemple Suisse*, Regio Basisliensis 24.

Ullman E.L., 1959, *Geografia transportu* (Geography of transport; Polish translation), [in:] PZLG, 1, IG PAN, Warszawa, pp. 1–30.

# PART III
# Comments

# Chapter 13

# The Expansion of Infrastructures as 'The Second Party Programme': A Look at the Bright and Dark Sides of Soviet Life

## Klaus Gestwa

*Insfrastruktura,* in the Russian language, is a borrowed word from Western academic and political linguistic usage of the 1960s. In 1972, it was officially incorporated by the Comprehensive Soviet Encyclopaedia (Bol'shaia Sovetskaia Entsiklopediia 1972: 1062–3; see also Ekonomicheskaia Entsiklopediia 1975: 61–2). The corresponding definition is nearly identical to those of the German and English terms. *Infrastruktura* refers to all materials, institutions and personnel facilities and establishments, which are available within a society based on the division of labour. These contribute significantly to precipitation of a maximum economic growth rate through interconnection and exchange. *Infrastruktura* is functionally understood as being regional and sectorial structural development, a joint task and a public institution. It is characterised by immense periods of construction, significant investments, extensive demand and diverse interrelations with other sectors. Until today, *infrastruktura* is a *terminus technicus* of political and journalistic usage in Russia. It is used much less frequently than in the German and English languages.

Although the term *infrastruktura* was not used as a signal or keyword in the Soviet Union until the 1960s, party leaders and planning experts were thoroughly familiar with the circumstances associated with the term. In debates about the forced industrialisation of the Soviet Union, these individuals spoke of the 'material-technical foundations' for the 'construction of socialism' and, later, for the 'transition to communism' (Novikov 1960). The technical infrastructures described by this term were integral to the self-conception of the Soviet party state. For the basic configuration of communal life, these served for networking purposes. As 'first-order societal integration media' (Laak 2001: 368), they were implemented as mediating entities, between the state and the economy, rule and daily life, culture and nature, urban and rural, and centre and periphery.

The following contribution describes, in its first section, how the creation of extensive and efficient energy, transport and communication systems served as a precondition for the integrative and developmental capacity of a growth-orientated Soviet economy. The backwardness remaining of Tsarist Russia was to be overcome. A new Soviet state was to appear on the passing lanes of history with maximal societal acceleration, in order to catch up the developments of the modern industrial age. Moscow's social technocrats used modern supply and disposal systems as governing instruments, thereby including the inhabitants of the first socialist state in the Soviet social experiment. The modern 'networks of power' (Hughes 1983) served to conquer time and space by way of an interior technological colonisation. The diverse regions of the red empire, many of which were economically and administratively barely developed, had to be connected with one another. Infrastructures were used to overcome boundaries and to enforce a standardised lifestyle, even in the most distant corners of the country.

The Soviet infrastructures, however, did not function solely due to the great promises of prestige and solidarity, or of prosperity and progress. They did not simply represent the impressively-staged success story of the Soviet state. The second section of the contribution presents a look behind the scenes of envious self-representation, thereby revealing the ruthless and neglected aspects of Soviet civilisation. The history of Soviet infrastructure is characterised by a fatal mixture of revolutionary impatience, economic intemperance and administrative disorganisation. The consequences of brutal *social engineering* are emphasised. It will become clear that the infrastructural expansion was accompanied by rigorous exclusionary mechanisms. In the perestroika period, the ailing state of Soviet infrastructures mirrored the miserable situation of the entire country. These bitter, increasingly encompassing realisations accelerated threatening disintegrating processes, which finally resulted in the collapse of the Soviet empire.

## 1. 'Always Look on the Bright Side of Life': Soviet Infrastructures and the Road to Communism

Lenin, himself, eloquently stated that technological infrastructures had a central function within the Soviet state as instruments of economic and political integration. In December of 1920, amidst the turmoil of the Russian Civil War, he proclaimed his legendary slogan: 'Communism is Soviet power plus the electrification of the whole country.' Lenin elevated his electrification plan, which entered into Soviet history under the acronym GOELRO, into the position of 'the second party programme'. Thereby, he assigned infrastructure planning the same level of importance as the ideological guidelines and premises formulated within the party programme. 'Only when the economy has been electrified, and industry, agriculture and transport have been placed on the technical basis of modern large-scale industry, only then shall we be fully victorious ... if Russia is covered with a dense network of electric power stations and powerful technical installations, our communist economic development will become a model for a future socialist Europe and Asia' (Lenin 1963: 160–61).

For Lenin and his successors, infrastructures were *incunabula* of progress. They were indicators of the capacity and level of development of the new Soviet power. If power is understood as being the capability to transform given conditions through independent decision-making and action, then the Soviet party state gained great power by way of infrastructural development. With ambitious projects, the backward Russian agricultural landscape was transformed, over a short period of time, into a leading industrial power and superpower. Stalin brought his ruthless actions and transformative power to a point: he found Russia with a wooden plough and left it with a nuclear reactor (Josephson 2000). Similarly, Soviet propaganda, following Stalin's death, emphasised the 'great leap from Mužik to Sputnik' (i.e. the transformation of a Russian peasant to a Soviet cosmonaut). This was a mighty success story, which was impressively embodied by Yuri Gagarin (Gestwa 2009a, Jenks 2012).

With Soviet infrastructure, an enormous mobilisation and directional power unfolded. For the realisation of important infrastructural developments, such as railways, canals and power stations, those responsible mobilised an army of labourers, modern machinery and mountains of construction materials. Supplying industries emerged just as suddenly as did new cities, which appeared all across the Soviet empire and even within the polar circle. The geography of Soviet economics and settlement took on a fundamentally new appearance.

With the construction of infrastructures, the population itself came into motion. The Soviet Union proved to be a 'quicksand society' (Lewin 1985), especially in the 1930s. The drifting social sand of Stalinism, however, did not disperse at random. Rather, it dispersed along new infrastructural arteries. Infrastructures channelled migratory and material currents by determining routes and destinations.

The Turkistan-Siberian Railway, which was known as the 'Turksib', serves as a vivid example. This continuous railway stretched from the western Siberian city of Novosibirsk, passed through Alma Ata and ended finally in Tashkent and Frunze. It was built between 1926 and 1931. A prestige project, it enabled the party leadership to emphasise its ambitions to venture the building of a bridge between Europe and Asia. As the first Stalinist railway project, its intent was to achieve unification with the Muslim periphery through the construction of modern transport and communication networks. Thereby, the Central Asian regions of the Soviet Union, Turkmenistan, Kazakhstan, Uzbekistan, Tajikistan and Kyrgyzstan, could be economically and politically integrated into the Soviet empire. Along the Turksib, the revolutionary locomotive would continue along to Central Asia. Cotton, which was being increasingly harvested, would make its way to the textile factories in Europe along the new transport route. This promoted the integration of the Soviet economy. Additionally, new industrial and mining sites were established along the railway. The Turksib came to be the transport backbone of Soviet Central Asia (Payne 2001).

At the time at which construction of the Turksib began, 3.5 million people lived in Kazakhstan. Among them were only 3,000 so-called proletarians, most of which lived as nomads. Their existence was acutely threatened by the influx of farmers from the European regions of the Soviet Union and by the brutal agricultural collectivisation of the 1930s. The Kazakhs were positioned at the chasm of the modern Soviet era, which brutally made its way across Central Asia. The Turksib, with its construction sites and later with its railway stations, came to be an important employer for the Kazakhs. Many accomplished the leap from the pressing poverty of nomadic existence into the modern industrial world. The Kazakh working class began to take form. Infrastructural construction, such as that of the Turksib, contributed to the urbanisation, industrialisation and proletarianisation of the rural population. It also led to the massive immigration of European professionals and labourers. A strong Russian minority established itself in the Central Asian regions of the Soviet Union. Between 1926 and 1939, its percentage of the population in Kyrgyzstan and Kazakhstan doubled from 21 to 40 per cent. The party leadership understood the modern spaces, created by the construction and expansion of infrastructures, to be important melting pots. There, traditions and lifestyles could melt together in order to form new social identities. Soviets with diverse social and ethnic backgrounds enthusiastically bonded with one another. Then, they would grow together and become a unified Soviet people (Gestwa 2005). Upon construction of the first, pompous underground railway lines in Moscow in the mid 1930s, the propaganda declared loudly and with pride: 'The whole country builds the Metro' (Neutatz 2001: 509). The large construction site was considered a 'place symbolic of friendship among peoples' (Zarevo nad Volgoj 1959: 265–6). The unified Soviet people (*Sovetskii narod*), which had come together through infrastructure and large development projects, sent a message to the world: 'It showed them, how peoples of the most diverse backgrounds can peacefully […] live together and create for a common benefit' (Wunderwerke unserer Epoche 1952: 33). However, the vision of a harmonious Soviet people remained more a dream than a reality, due to continuing ethnic and social tensions. Still, for the self-representation of the Soviet party state, it was an indispensable vision. Its effectiveness as a political myth continued to unfold, even until today in the post-Soviet era (Martin 2001).

The mobilisation power of infrastructures was based on their power to define and persuade. Every important infrastructural measure went hand-in-hand with extensive propaganda campaigns. These were intended to inspire, carry away and draw in the Soviet people in respect to the establishment of socialism and communism. *Social engineering* was, thereafter, closely connected to *soul engineering* (Gestwa 2009b). Soviet infrastructures were more than merely economic foundations. As sources of enthusiasm, they served as events of purpose and arsenals of legitimation. A never-ending sense of departure and an optimistic attitude towards life, despite all hardships, were integral to the construction and expansion of infrastructures. In the 1970s, a Siberian party official emphasised how important these staged, cultural scenes were for the Soviet economics of salvation. Soberly, he explained the unswerving necessity for new, expensive hydraulic power stations and railways. Without such great undertakings, 'no idealism would remain in this country' (Smith 1976: 420–21). This confession emphasises that the cult of grandiose infrastructural projects took on an indispensable pivotal position between the hopes of the population and the expectations of the regime. Large infrastructural projects, greatly staged, were to convince the Soviet people and prepare them for the transition to communism. The paradise of tomorrow was gauged with these projects. The simplicity of the reconfiguration, exemplarily presented as a space of the future, is suggestive of the idea of a simple restructuring of the whole country and, thereby, of reality.

Even the Soviet Union of the Brežnev era, which, supposedly, was orientated towards stabilisation, built upon the Soviet tradition of permanent departure and constant state of emergency. The Baikal-Amur Mainline (BAM), a project which was begun in 1974, is one example. The aging Kremlin leadership, under Leonid Brežnev, decided to set the unfinished project of a railway line to eastern Siberia on its political agenda anew. Through tremendous financial and societal exertions of force, it was to be completed. So-called territorial production complexes, meant to serve the expanding mining and industrial regions, were planned along the lines of the celebrated mainline. These were to stimulate the faltering Soviet economy and prepare its way into a 'light future'. This final, great Soviet mobilisation project relied entirely upon the socially integrative function and dynamic power of overwhelming infrastructure projects. The party leaders and propaganda strategists attributed a key societal function to the BAM. It was to become, once again, a 'symbol of an era' and an 'experience of a generation'. The new transportation mainline illuminated, as a lighthouse of progress, the light way into the soon-to-come third millennium. The lofty BAM myth was to suggest economic potential, sustainability and the willingness to expand, despite the fact that societal development had long lost its energy and momentum. The 'Wild East' of still-isolated Siberia served as a projection surface for conveying the promise of freedom, adventure and romance. The building labour force was considered, once more, to be a peacefully coexisting and cooperating 'Soviet people of a small scale' (Grützmacher 2012).

The power of seduction and integration, inherent to Soviet infrastructure, resulted by way of reward. It was founded not on idealistic motives, but on that which, in Soviet political jargon, was referred to as the 'principles of materialistic interest of the working people'. Those who became active in the establishment of socialism and communism were rewarded with promising earnings and career opportunities. Infrastructure provided for education and qualification opportunities and allowed millions of Soviets, as long as they possessed the relevant know-how, to climb up the social ladder. The Soviet elite, after the 1950s, consisted increasingly of party and state officials from agricultural and/or working backgrounds, who had been educated as engineers. They had been socialised by Stalin's great industrial and infrastructural projects, and had been politicised by World War II. Thus, as banner carriers of a technocratic ethos, they put their stamp on Soviet politics up until and even throughout the perestroika period (Bailes 1978, Beyrau 1993, Schattenberg 2002).

## 2. 'The Dark Side of the Moon': Soviet Infrastructures as Means of Repression, Exhaustion and Destruction

### 2.1 The GULag Economy: Infrastructure of Coercion and Inhumanity

Those who refused to participate in infrastructural construction during the Stalin era, or voiced doubts about its usefulness, risked being arrested or murdered. Between 1931 and 1956, countless railways, canals and power stations were built by an army of forced labourers. Additionally, these labourers, prisoners of the infamous GULag system, were also forced to construct roads, bridges, telephone lines, schools, universities and hospitals. Soviet infrastructures were built upon blood and violence. They emanated a power of intimidation, as well as a power of seduction, acting as terrible disciplinary agents. The suffering and plight of millions of forced labourers are just as much part of the history of Soviet infrastructure as are the celebrated records and breakthroughs (Khlevniuk 2004, Hedeler 2010, Barnes 2011).

As the first major construction project of the Stalinist administration, the White Sea–Baltic Sea Canal (*Belomorkanal*) gained a tragic fame. Even though the survey work still had not been finished, after seven months, the decree to begin with the construction of the 227-kilometer-long canal was issued in September of 1931. Large camp complexes appeared overnight and filled quickly. The total number of forced labourers reached 170,000. With only the simplest of tools, they were forced to move huge masses of earth and install harbour and lock facilities. The White Sea–Baltic Sea Canal, proudly named after Stalin, was opened in August of 1933 for shipping traffic. The responsible authorities calculated that the project had cost only 93 million Roubles, instead of the 400 million Roubles it had been estimated to cost in advance. However, no one spoke of the excessive haste, faulty planning and resulting problems. Stalin, himself, was not especially impressed with the waterway as he travelled along a stretch of the canal in August of 1933. To him, the canal seemed too narrow. Furthermore, it did not have the depth necessary to allow large ships to pass through (Ruder 1998, Morukov 2003). As a strategically important transport route, the White Sea–Baltic Sea Canal did not satisfy expectations. During the short period of navigation on the canal, few ships passed through its locks. Passenger and freight transport proceeded mostly along the neighbouring railway lines. 'Chosen in haste, born blindly' – taunted the author and dissident Aleksander Solzhenitsyn later (Solzhenitsyn 1978: 95).

The cultural programme supporting the construction of the Belomor Canal, especially, made history. In 1932, a ship carrying journalists, authors and other creative artists was chartered. On a journey into the far north, the passengers were to view the canal. In form of a large-scale media campaign, they propagandistically portrayed the construction of the canal not only as an ambitious waterway project, but also as an impressive re-education programme for criminals, 'parasites' and 'vermin'. The corresponding anthology, published also in English, included contributions of 36 famous Soviet writers. Today, it is considered to be an impressive testimony to the terror of the Stalinist regime (Belomorsko-baltijskij kanal 1934). Stalin, himself, proved to be very pleased by the political engagement of the Soviet writers and, from then on, hailed them as 'engineers of the human soul' (Westerman 2006). These descriptions expressed Soviet state's aspirations to re-structure the inner workings of the Soviet people, as well as to re-structure the societal and natural environment. 'By changing nature, we change ourselves' – with these words, Maxim Gor'kij indoctrinated the prisoners (Weiner 1995). The camp complexes along the Belomor Canal were referred to as 'improvement labour camps'. The brutal secret service officers and GULag bosses were celebrated as social educators. These individuals turned class enemies into true enthusiasts and new persons by way of forced labour and political indoctrination. In reality, their actions were degrading and striking.

## 2.2 The Concentration of Power: Centralisation and Exhaustion

Soviet infrastructure often served to administratively integrate peripheral areas, thereby making them economically accessible. Regional party leaders often acted publicly as vehement supporters of large infrastructure projects. Behind the scenes, they acted as lobbyists. However, this cannot deceive of the fact that decisions concerning the construction of infrastructures were mostly made in Moscow and that implementation processes were predominantly organised centrally. By and large, the cartels of experts in Moscow and Leningrad maintained full control of planning and construction activities. They succeeded in maintaining the trust and protection of party leaders, and used their influence to achieve institutional transformations to their own advantage. The infrastructural expansion allowed for the establishment of important central and planning agencies in Moscow and Leningrad. With their blueprints, these agencies determined Soviet investment policy. A new structural power emerged at the central hubs between planning and administration.

Khrushchev's reform policies of the late 1950s led to a decentralisation of political power under the banner of thawing and de-Stalinisation. In the following years, however, many changes were reversed and the power of the Moscow ministries was strengthened anew. Despite this, new economic centres, research institutes and planning institutes were established in the capital cities of the various Soviet states. Among them were important Russian regions, such as Siberia. These demonstrated their own initiative and presented plans for infrastructural expansion (Josephson 1997, Obertreis 2008). The political leadership of the Soviet states maintained the right to realise infrastructural projects costing less than 50 million Roubles, without the explicit agreement of authorities in Moscow. A 'process of republicanisation' began (Marc Elie 2013). The individual Soviet republics increasingly coordinated planning and construction activities. However, differences of power and competency continued to exist between Moscow and the peripheral regions. Many infrastructure projects still depended upon the expertise, competency and resources of the central institutes in Moscow. Furthermore, important decisions could hardly be made without the backing of the leadership in Moscow. Although the Soviet republics achieved an increase in political influence and in planning and administrative participation, the infrastructural particularities of individual regions remained categorically excluded (Lehmann 2012).

Soviet infrastructures were almost always oriented towards an urban-industrial centre. They concentrated powers. As the giant hydraulic power station on the Volga River by Kujbyšev and Stalingrad commenced its operations in the 1950s, the majority of the electric energy produced travelled directly to Moscow, through power lines of up to 900km in length. The Volga region itself did not greatly profit. Supplying the centre always had priority over the nationwide usage of energy infrastructures. The interests of the countryside and the agricultural sector stood far behind the demands of cities and industries. Thereby, social hierarchies and differing positions of power manifested themselves within Soviet infrastructures. As a result, many Soviet citizens, especially in rural areas, had to wait for the inclusion into the modern supply and disposal systems promised by Lenin for a very long time. Powerless, they had to endure their systematic neglect. To them, the industrial Soviet civilisation revealed itself as an exhausted modern entity, which had focused sparse resources in other areas. For this reason, the Soviet Union, in the 1960s, joked that Lenin's 1920 slogan needed to be re-stated in mathematical terms. Based on the everyday experiences of millions of Soviet citizens, a better slogan would be: 'Soviet power is communism minus the electrification of the whole country.' In 1965, the newly-appointed Secretary General Leonid Brezhnev announced, once again, a special programme that would last several years and supply all rural settlements permanently and sufficiently with electricity (Gestwa 2010: 224–35).

The rural residents were, most definitely, among the losers of Soviet infrastructural expansion, as they did not reap equal benefits. The transport routes to agrarian settlements were often in such a poor condition that much of the harvest spoiled in the fields. The underdeveloped agricultural sector continued to be the Achilles heel of the Soviet economy (Hanson 2003: 74–80, 149–54).

As mechanisms of integration, infrastructures were to bring together the urban centres and rural periphery, and minimise cultural differences. Without a doubt, this goal was not achieved. Infrastructures did not only illustrate the chasm between city and country. Often, they even increased it. However, there was one interesting exception. When television was introduced as a new medium in the Soviet Union, it was, within a relatively short period of time and despite initial production problems, able to reach the entire Soviet population. By the start of the 1970s, more than 90 per cent of all Soviet households were equipped with a television set. Television was an essential channel for propaganda. The Soviet party state promoted this form of vertical communication, especially, in order to send its messages of power to its citizens, even in the most distant corners of the empire (Roth Ey 2011).

The development of horizontal communication infrastructures, which would have accelerated the exchange of information among the people, lagged behind

Russians. The success of the village writers demonstrates that, in the Soviet Union of the 1960s and after, the alienation potential of modern infrastructures was more present in public debates than was their potential for integration (Janickij 1992, Parthé 1992, Kochanek 1999).

The environmental pollution caused by Soviet infrastructures is another aspect of the dark side of destruction. When the Soviet Union collapsed, nearly one-third of its area was considered to be an 'ecological disaster region'. The responsible politicians, planners and managers had clearly invested much more in production and distribution than in purification and disposal. Filtration plants, wastewater treatment plants and security measures had been low on the political agenda. The maintenance and service of infrastructures were also often insufficient. In addition, the authorities responsible for operational safety took unnecessary risks in order to realise their plans. The fatal consequence was that Soviet infrastructures polluted their surroundings immensely, even as a result of their normal, problem-free operations. In addition, countless failures and accidents resulted in the leakage of large amounts of pollutants. In the early 1990s, the industrial cities of Magnitogorsk, Noril'sk and Bratsk were high on the list of the most polluted cities in the world. Not only here did the industrial Soviet civilisation reveal itself as a negligent modern state, which carelessly wasted resources, spoiled landscapes and criminally threatened human health. The nuclear disaster in Chernobyl on April 26, 1986, was the tragic climax of the history of Soviet catastrophes. The drying of the Aral Sea, as a result of excessive irrigation, is another catastrophic example of falsely planned infrastructure projects. Towards the late 1980s, the Soviet press, enabled by glasnost policies to discuss problems and shortcomings, revealed the shaky foundation of Soviet modernity. This shaky foundation resulted from the massive risks taken on behalf of infrastructural expansion, which could hardly be controlled or minimised (Dawson 1996, Gestwa 2003).

With its worn out and often extremely risky infrastructure, the Soviet state no longer appeared to be a reliable guarantor for security. More so, it revealed itself as being organised irresponsibility with a high potential for catastrophe. The evident fragility of ideological promises was revealed to Soviet citizens not only in political and cultural debates, but also in the repeated experience of the unpredictability of life and in the heavy burden of pressing daily problems. The term *progress* was once the most confident expression of the unwavering planning optimism in Moscow. Now, it decayed in the delirium of an increasing loss of confidence of a culture of fear, indignation and outrage. Since the end of the 1980s, a continuing disaster scenario darkened the political climate. A spiral of mistrust degraded the institutions of the party state. The infrastructural power of integration, largely staged by Soviet propaganda, had long been used up (Altrichter 1989, Gestwa 2011).

## 2.5 'A Model for a Future Socialist Europe and Asia':
## The Export of Soviet Infrastructures

In his famous speech on electrification, in 1920, Lenin predicted that the Soviet Union, with its 'powerful technical installations ... [would] become a model for a future socialist Europe and Asia (Lenin 1963: 161)'. After achieving the status of 'superpower', following World War II, the Soviet Union began to export both its societal model and its infrastructures. Beginning in the late 1950s, the unforgiving system rivalry of the Cold War increasingly determined which superpower could win over the global public with its model of modernity. The most well-known, international Soviet project was the monumental Aswan Dam in Egypt. The Soviet Union took over the planning and financing of the project in 1958, after Western companies and banks had backed out on their agreements following the Suez Crisis of 1956. Soviet propaganda declared the hydropower giant, officially inaugurated in 1971, to be a 'symbol of the friendship between the Soviet people and the Eastern nations' (Nesteruk 1965: 45). The political symbolism of the Aswan Dam was rooted in the promising message sent by the Soviet Union to all of the underdeveloped and suffering regions of the world. It was a symbol for that which Moscow planned to do for those nations prepared to shake of the yoke of colonialism and affiliate themselves with the 'camp of socialism and peace'. The sonorous propaganda, however, concealed the countless problems that arose as a result of the technical cooperation and political relationship between Moscow and Cairo. Furthermore, the Soviet Union was exporting, through infrastructure, its own form of destructive progress. Similar to the situation of the Volga and Siberian Rivers, countless historical settlements, historical sites and rich agricultural lands along the Nile River disappeared with the completion of the Aswan Dam. The dammed Nile River and its neighbours soon suffered from the environmental consequences, which had been deliberately overlooked by experts during the planning and building phases. For many of the affected, the dam was more a curse than a blessing (Bishop 1997, Biswas/Tortajada 2004).

The extent to which planners and politicians in Moscow trusted in infrastructures as powerful instruments of political and economic integration is impressively evident in the history of the Eastern Bloc. After 1945, once the Soviet realm of power extended all the way to the Elbe River, the focus was for the so-called Eastern Bloc to achieve internal cohesion and to be animated with an enthusiastic, cooperative spirit. However, the Council for Mutual Economic Assistance (COMECON) – founded in 1949 as a socialist, Eastern 'replacement Europe' and often reformed – was never able to realise the growth, integrative and emancipatory potential of the rapidly developing Western European economic and values community. Ambitious cooperative efforts failed repeatedly as a result of national egoism or acute weaknesses of coordination. Often, the potential of the Eastern European planned economies was not enough to complete these projects. Furthermore, the party leadership in Moscow did not prove itself willing to ease up on its hegemonic demands opposite its allies (Thum 2004, Dangerfield 2010).

For a time, it was even easier to travel through the Iron Curtain, to the West, as it was to cross the isolated borders to neighbouring states within the Eastern Bloc (Trutkowski 2011; cf. with the contribution by Tomasz Komornicki, also in this reader).

Irrespective of these problems, the party leadership of the Eastern Bloc never stopped believing that the Eastern European allies would find one another and become a 'socialist community' by way of large infrastructure projects. They trusted in the cohesive power of common work efforts. Large construction sites, as showplaces of international friendship, were to bring people together. Thereby, the historical power of 'internationalism' could become a tangible experience. Transnational infrastructures, as a result of the forced circulation of people and commodities, did indeed lead to trans-border integration and networking. Prominent examples of Eastern European networking are evident, especially in the energy sector as a result of pipeline construction. The Druzhba oil pipeline was built between 1959 and 1964. In 1975, the construction of the Soyuz and Bratstvo natural gas pipelines began. The names of the three pipelines – meaning *friendship, unity,* and *brotherhood* – clearly reflected the political aspirations. The construction of this pipeline network, which stretches across thousands of kilometres – from Soviet sites far beyond the Volga and Ural regions all the way to the centre of Europe –, was a joint venture of the Eastern Bloc. During COMECON meetings, carefully drafted, international agreements were signed. All European COMECON members participated in the construction of the pipelines. They did so by taking on the completion of sections of pipeline in return for free or inexpensive delivery of oil and natural gas. The party press spoke euphorically of the 'building project of the century'. The joint venture was celebrated as a shining example of the economic assistance given among states of the Eastern Bloc and of their cooperative spirit in general. 'Friendship among nations in action' – so read the popular slogan (Eggers 1978, Bencard/Taubert 1985, Obuchoff 2012). In the 1970s and 1980s, for example, the GDR sent more than 30,000 labourers to the Ukraine and Belarus to finish sections of pipeline in those regions. The youth organisation of the Communist Party recruited young people, especially, for this task. They were promised an organised adventure in the 'land of Lenin', which would be full of excitement and bring out the best in those who took part (Best 2008 and 2010, Prochnow 2010).

Beyond its propagandistic and political value, pipeline construction had immense economic benefits. Without the delivery of energy from the Soviet Union, which was enabled through the construction of the pipelines, the states of the Eastern Bloc would hardly have been able to achieve economic growth. The trans-border pipe systems formed infrastructural veins, through which power currents kept the Eastern Bloc alive.

In addition, the transcontinental pipelines are a visible example of how the Soviet Union and the Eastern Bloc wrote the history of European interactions across the blocs. In the 1970s and 1980s, the pipes for the pipeline construction originated, to a large extent, from West Germany. They were delivered by the company Mannesmann, which aimed for large profits with its trade of natural gas pipes to the Soviet Union. Furthermore, these pipeline networks were the basis of rising Soviet energy exports to Western Europe. Until today, the pipeline network, built as a joint COMECON venture, is the infrastructural backbone of the entire European energy system (Gustafson 1983, Crovitz 1984, Jettleson 1986, Rudolph 2004, Götz 2004, Balmeceda 2004).

In the 1980s, even the economically beneficial pipelines gained a negative reputation. However, the much-praised 'proletarian internationalism' did not reach a standstill on its own. Foreign workers were often legally inferior and discriminated (Jaješniak-Quast 2005). The pipelines, often brutally constructed within fragile ecosystems, passed through the pastures of many herds. Reindeer and other animals were isolated from their grazing areas. This led to the destruction of the lifestyles and economic traditions of indigenous peoples (Dahlmann 2009: 270–97; Bruno 2011). Oil and natural gas leaked regularly from the pipelines, which damaged the environment and climate. The gas pipeline disaster in the Ural region, which occurred on the night of June 3rd, 1989, caused an outrage. A large amount of natural gas had leaked from a gas pipeline running parallel to the Trans-Siberian Railway route. This gas caught fire, which sparked a large explosion just as two long-distance trains were passing one another nearby. More than 600 people died. Among them were many children, who were on holiday to the Black Sea. The disaster made very clear how catastrophic the state of Soviet infrastructures was. The gas pipelines were no longer symbolic for the friendship among nations, but rather for the general lack of responsibility. Out of promise emerged disaster (Altrichter 2009: 250–53).

This oppressing realisation spread throughout the Eastern Bloc. Even in the Soviet satellite states, people increasingly lost confidence in their infrastructures and in the ruling party leadership. The imported Soviet nuclear power stations, of the Chernobyl-type, caused the most concern. Many of these had already commenced with operations, thereby turning the nuclear risk into a communal problem. A transnational hazard-zone emerged as a result, within which production, service and emergency measures often did not follow the appropriate regulations.

'To learn from the Soviet Union means to learn to be victorious' – this propaganda slogan, common in the GDR, had long been reversed. With the purchase of Soviet reactors, people far beyond Soviet borders found themselves to be nuclear hostages. Beginning in the late 1980s, especially in the GDR and Czechoslovakia, people protested vehemently against the production of nuclear power. In doing so, they increasingly denied the socialist party state their faith and following (Abele 2000, Arndt 2011). This nuclear-powered communism, once a bubbling spring of optimism, had turned out to be the error of post-war, socialist modernity. The legitimacy of Soviet economics gave way along with the trust in Soviet infrastructures. The wearing down of infrastructures and the loss of integrative power were two interconnected processes, which ultimately resulted not only in the collapse of the Soviet empire, but also of the entire Eastern Bloc.

## 3. Conclusion: The Dynamics of Integration and Disintegration

The founders of the Soviet Union, as early as 1920, declared the construction of infrastructures to be the 'second party programme'. The foundations of Soviet power, therefore, were not alone the powerful party and state institutions of Moscow. The infrastructural networks, which crossed and connected the Soviet empire, were just as important. The expansion of infrastructure is, without a doubt, a central aspect of Soviet history. It mobilised an immense mass of people and resources, created new structures of time and space, and produced overwhelming images of a light future and of blossoming landscapes. These came to be an integral part of the Soviet perception of reality. As 'networks of power', infrastructures were not only politically and economically, but also socially and culturally integrative. The unified 'Soviet people' and 'socialist community' of Eastern European allies were imaginary constructs of Soviet propaganda, meant to become reality in form of trans-border infrastructures. Undoubtedly, the construction and expansion of modern infrastructures contributed to the increasing power of the party leadership in Moscow. A guarantee of continuance developed and continued to be effective, despite defects and shortages, until the dark side of the Soviet infrastructures became the object of public discourse. Due to the disillusioning balance sheet of destruction and its enormous social vulnerability the first socialist state, which had gained momentum as a superpower after 1945, ultimately proved to be a run-down and shabby modernity. It had long forfeited its creative powers. Scruffy infrastructures led to increasingly disintegrative results, rather than to the expected integrative effects. Mikhail Gorbachev attempted, beginning in 1985 with his perestroika policies, the overdue modernisation of the Soviet modernity. His ambitious reform policies failed, however, opposite a society which had long lost its faith in a better, socialist world.

Today, the abandoned, 30-square-kilometre zone around the nuclear reactor in Chernobyl seems like a contaminated mausoleum of past dreams of a better life. The impressive technical and social infrastructure which remains has become increasingly overgrown (Kostin 2006, Krementschouk 2011). There, the powerlessness of the party state becomes especially visible.

These observations correspond to the realisation that Soviet infrastructures, despite all current processes of decay and border demarcation, serve anew as media for integration. Transport and energy networks, as well as telephone, gas and power lines, are cohesive elements for the post-Soviet sphere. In some cases, such as that of the Turksib Railway, they have come to be objects of nostalgic glorification. The successes and potential of Soviet infrastructures are paid attention once more. The revealing ambivalence of the history of Soviet infrastructures is part of today's post-Soviet reality, finding it difficult to recede into the past.

## References

Abele, J. 2000. *Kernkraft in der DDR. Zwischen nationaler Industriepolitik und sozialistischer Zusammenarbeit 1963–1990*. Dresden: Hannah-Arendt-Institut für Totalitarismusforschung.

Altrichter, H. 2009. *Russland 1989. Der Untergang des sowjetischen Imperiums.* München: C.H. Beck.

Arndt, M. 2011. *Tschernobyl: Auswirkungen des Reaktorunfalls auf die Bundesrepublik Deutschland und in die DDR.* Erfurt: Landeszentrale für politische Bildung Thüringen.

Bailes, K.E. 1978. *Technology and Society under Lenin and Stalin: Origins of the Soviet Technical Intelligentsia, 1917–1941.* Princeton, NJ: Princeton University Press.

Balmaceda, M.M. 2004. Der Weg in die Abhängigkeit: Ostmitteleuropa am Energietopf der UdSSR. *Osteuropa* 54, 162–79.

Barnes, S.A. 2011. *Death or Redemption: The Gulag and the Shaping of Soviet Society.* Princeton and Oxford: Oxford University Press.

Belge, B. und Gestwa, K. 2009. Wetterkrieg und Klimawandel: Meteorologie im Kalten Krieg. *Osteuropa* 59, 15–42.

*Belomorsko-baltijskij kanal* imeni Stalina 1934. Istorija stroitel'stva, Moskva. (*Belomor. An Account of the Construction of the New New Canal between the White Sea and the Baltic Sea*). New York 1935)

Bencard, Th./Taubert, K. (ed.) 1985. *Am Bauwerk des Jahrhunderts: Erlebnisse vom Zentralen Jugendobjekt 'Erdgastrasse' der Freien Deutschen Jugend.* Berlin: Neues Leben.

Best, U. 2008. Die anderen Räume des Sozialismus. Internationale Baustellen in der Sowjetunion und ihre Erinnerung. *Kakanien Revisited* [Online Journal]. Available at: http://www.kakanien.ac.at/beitr/emerg/UBest1.pdf [12 November 2013].

Best, U. 2010. Arbeit, Internationalismus, und Energie. Zukunftsvisionen in den Gaspipelineprojekten des RGW, in: *Zukunftsvorstellungen und staatliche Planung im Sozialismus. Die Tschechoslowakei im ostmitteleuropäischen Kontext 1945–1989*, edited by Schulze Wessel M./Brenner, Ch. München: Oldenbourg, 137–58.

Beyrau, D. 1993. *Intelligenz und Dissens: Die russischen Bildungsschichten in der Sowjetunion 1917 bis 1985*, Göttingen: Vandenhoeck & Ruprecht.

Bishop, E. 1997. *Egyptian Engineers and Soviet Specialists at the Aswan High Dam.* Ph.D. Dissertation, University of Chicago.

Biswas, A.K. und Tortajada C. 2004. *Hydropolitics and Impacts of the High Aswan Dam.* Mexico City.

*Bol'shaia Sovetskaia Entsiklopediia* 1972. Vol. 10, 3rd Edition, Moskva.

Bruno, A.R. 2011. *Making Nature Modern: Economic Transformation and the Environment in the Soviet North.* Ph.D. Dissertation, University of Illinois at Urbana-Champaign.

Crovitz, G. 1984. Europe's Siberian Gas Pipeline. Economic Lessons and Strategic Implications. London.

Dahlmann, D. 2009. *Sibirien vom 16. Jahrhundert bis zur Gegenwart.* Paderborn: Schöningh.

Dangerfield, M. 2010. Sozialistische ökonomische Integration: Der Rat für gegenseitige Wirtschaftshilfe (RGW), in Ökonomie im Kalten Krieg edited by B. Greiner et al. Hamburg: Hamburger Edition, 348–69.

Dawson, J.E. 1996. *Eco-Nationalism: Anti-Nuclear Activism and National Identity in Russia, Lithuania, and Ukraine.* Durham and London: Duke University Press.

Eggers, G. et al. 1978. *Abenteuer Trasse. Erlebnisse und Beobachtungen.* Berlin: Neues Leben.

Elie, M. 2013. Coping with the 'Black Dragon'. Mudflow hazard and the Controversy over the Medeo Dam in Kazakhstan, 1958–1966. *Kritika* 14 (2), 313–42.

*Ekonomicheskaia Entsiklopediia* 1975. Vol. 2, Moskva.

Gestwa, K. 2001. Infrastrukturen – Kommunikation, in *Handbuch der Geschichte Russlands. Bd. 5: 1945–1991. Vom Ende des Zweiten Weltkriegs bis zum Zusammenbruch der Sowjetunion,* edited by S. Plaggenborg. Stuttgart, 1142–52.

Gestwa, K. 2003. Ökologischer Notstand und sozialer Protest: Der umwelthistorische Blick auf die Reformunfähigkeit und den Zerfall der Sowjetunion. *Archiv für Sozialgeschichte* 43, 349–84.

Gestwa, K. 2004. 'Energetische Brücken' und 'Klimafabriken': Das energetische Weltbild der Sowjetunion. *Osteuropa* 54, 14–38.

Gestwa, K. 2005. Technologische Kolonisation und die Konstruktion des Sowjetvolkes: Die Schau- und Bauplätze der stalinistischen Moderne als Zukunftsräume, Erinnerungsorte und Handlungsfelder, in *Mental Maps – Raum – Erinnerung: Kulturwissenschaftliche Zugänge zum Verhältnis von Raum und Erinnerung,* edited by D. Geilsdorf. Münster: Lit, 73–115.

Gestwa, K. 2007. Auf Wasser und Blut gebaut: Der hydrotechnische Archipel GULag, 1931–1958. *Osteuropa* 57, 239–66.

Gestwa, K. 2009a. 'Kolumbus des Kosmos': Der Kult um Jurij Gagarin. *Osteuropa* 59, 121–52.

Gestwa, K. 2009b. Social und Soul Engineering unter Stalin und Chruschtschow 1928–1964, in *Die Ordnung der Moderne: Social Engineering im 20. Jahrhundert,* edited by T. Etzemüller. Bielefeld: Transcript, 241–78.

Gestwa, K. 2010. *Die 'Stalinschen Großbauten des Kommunismus': Sowjetische Technik- und Umweltgeschichte 1948–1964.* München: Oldenbourg.

Gestwa, K. 2011. Sicherheit in der Sowjetunion 1988/89: Perestrojka als missglückter Tanz auf dem zivilisatorischen Vulkan. in *Schlüsseljahre: Zentrale Konstellationen der mittel- und osteuropäischen Geschichte,* edited by M. Stadelmann and L. Antipow. Stuttgart: Franz Steiner, 449–68.

Götz, R. 2004. Pipelinepolitik: Wege für Russlands Erdöl und Erdgas. *Osteuropa* 54, 111–30.

Grützmacher, J. 2012. *Die Baikal-Amur-Magistrale: Vom stalinistischen Lager zum Mobilisierungsprojekt unter Brežnev*. München: Oldenbourg.

Gustafson, T. 1983. *The Soviet Gas Campaign. Politics and Policy in Soviet Decisonmaking*. Santa Monica: Rand Corp.

Hanson P. 2003. *The Rise and Fall of the Soviet Economy*. London: Pearson.

Hedeler, W. 2010. Ökonomik des Terrors: Zur Organisationsgeschichte des Gulag 1939 bis 1960. Hannover: Offizin.

Hirsch, F. 2005. *Empire of Nations: Ethnographic Knowledge and the Making of the Soviet Union*. Ithaca: Cornell University Press.

Hughes T.P. 1983. *Networks of Power: Electrification in Western Society 1880–1930*. Baltimore: Johns Hopkins University Press.

Jajeśniak-Quast, D. 2005. 'Proletarische Internationalität' ohne Gleichheit: Ausländische Arbeitskräfte in ausgewählten sozialistischen Großbetrieben in *Ankunft – Alltag – Ausreise. Migration und interkulturelle Begegnung in der DDR-Gesellschaft,* edited by C.T. Müller and P.G. Poutrus. Köln: Böhlau, 267–94.

Jajeśniak-Quast, D. 2009. Nowa Huta – Eisenhüttenstadt. Wirtschaftliche und zwischenmenschliche Kontakte auf Betriebsebene. Histoire Croisée oder 'Freundschaftstheater'?, *Studia Historiae Oeconomicae* 27, 295–305.

Janickij, O. 1992. Die ökologische Bewegung in Rußland – Versuch eines Soziogramms. *Osteuropa* 42, 694–702.

Jenks, A.J. 2012. *The Cosmonaut Who Couldn't Stop Smiling: The Life and Legend of Yuri Gagarin*. DeKalb: Northern Illinois University Press.

Jettleson, B.W. 1986. Pipeline Politics. The Complex Political Economy of East-West Energy Trade. Ithaca: Cornell University Press.

Josephson, P.R. 1997. *New Atlantis Revisited: Akademgorodok, the Siberian City of Science*. Princeton, NJ: Princeton University Press.

Josephson, P.R. 2000. *Red Atom: Russia's Nuclear Power Program from Stalin to Today.* New York: University of Pittsburgh.

Khlevniuk, O. 2004. *The History of the Gulag: From Collectivization to the Great Terror*. New Haven, CT: Yale University Press.

Kochanek, H. 1999. *Die russisch-nationale Rechte von 1968 bis zum Ende der Sowjetunion: Eine Diskursanalyse*. Stuttgart: Franz Steiner.

Kostin, I. 2005. *Tschernobyl: Nahaufnahme*. München: Kunstmann.

Krementschouk, A. 2011. *Chernobyl Zone* (I). Heidelberg: Kehrer.

Laak, D. v. 2001. Infra-Strukturgeschichte. *Geschichte und Gesellschaft,* 27, 367–93.

Lehmann, M. 2012. *Eine sowjetische Nation. Nationale Sozialismusinterpretationen in Armenien seit 1945*. Frankfurt a.M: Campus.

Lenin, V.L. 1963. *Polnoe sobranie sochinenii*. Vol. 42. 5. Edition. Moskva.

Lewin, M. 1985. *The Making of the Soviet System*. London: New Press.

Martin, T. 2001. *The Affirmative Action Empire: Nations and Nationalities in the Soviet Union 1923–1939*. Ithaca: Cornell University Press.

Morukov, M. 2003. The White Sea-Baltic Canal in: *The Economics of Forced Labor. The Soviet Gulag,* edited by P.R. Gregory und V. Lazarev. Stanford CA: Hoover, 151–62.

Nesteruk, F. Ja. 1963. *Razvitie gidroenergetiki SSSR.* Moskva.

Neutatz, D. 2001. *Die Moskauer Metro: Von den ersten Plänen bis zur Großbaustelle des Stalinismus.* Köln: Böhlau.

Novikov, I.T. 1960. *Elektrifikacija SSSR – važnejšnyj faktor sozdanija material'no-techničeskoj bazy kommunizma.* Moskva.

Obertreis, J. 2008. Der 'Angriff auf die Wüste' in Zentralasien: Zur Umweltgeschichte der Sowjetunion. *Osteuropa* 58, 37–65.

Obuchoff, H. u.a. 2012. Die Trasse. Ein Jahrhundertbau in Bildern und Geschichten. Berlin: Das Neue Berlin.

Palmer, S.W. 2006. *Dictatorship of the Air: Aviation Culture and the Fate of Modern Russia.* Cambridge and New York: Cambridge University Press.

Parthé, K.F. 1992. *Russian Village Prose: The Radiant Past.* Princeton, NJ: Princeton University Press.

Payne, M.J. 2001. *Stalin's Railroad: Turksib and the Buildung of Socialism.* Pittsburgh, PA: Pittsburgh University Press.

Prochnow, J. 2010. 'West Germans Don't Even Know about it': An Analysis of Narrations by East German COMECON Pipeline Workers after the Fall of the Wall. *Australian and New Zealand Journal of European Studies*, 2 (1), 16–34.

Roth-Ey, K. 2011. *Moscow prime time: how the Soviet Union built the media empire that lost the cultural Cold War.* Ithaca, NY: Cornell University Press.

Ruder, C.A. 1998. *Making History for Stalin: The Story of Belomor Canal.* Gainesville: University Press of Florida.

Rudolph, K. 2004. *Wirtschaftsdiplomatie im Kalten Krieg: Die Ostpolitik der westdeutschen Großindustrie 1945 – 1991.* Frankfurt: Campus.

Schattenberg, S. 2002. *Stalins Ingenieure: Lebenswelten zwischen Technik und Terror in den 1930er Jahren.* München: Oldenbourg.

Smith, H. 1976. *Die Russen.* Bern und München: Knaur.

Solzhenitsyn, A. 1978. *Der Archipel GULAG 2: Arbeit und Ausrottung, Seele und Stacheldraht.* Reinbek: Rowohlt.

Thum, G. 2004. 'Europa' im Ostblock: Weiße Flecken in der Geschichte der europäischen Integration. *Zeithistorische Forschungen/Studies in Contemporary History* [Journal], 1 (3). Available at: URL: http://www.zeithistorische-forschungen.de/16126041-Thum-3-2004 [12 November 2013].

Trutkowski, D. 2011. *Der geteilte Ostblock. Die Grenzen der SBZ/DDT zu Polen und der Tschechoslowakei.* Köln: Böhlau.

Weiner, D.R. 1995. Man of Plastic: Gor'kii's Vision of Humans in Nature. *Soviet and Post-Soviet Review* 22, 65–88.

Weiner, D.R. 1999. *A Little Corner of Freedom: Russian Nature Protection from Stalin to Gorbachev.* Berkeley: University of California Press.

Welitsch, W. 1952. *Die Eroberung der Wüste Karakum.* Leipzig: Reclam.

Westerman, F. 2003. *Ingenieure der Seele: Schriftsteller unter Stalin – Eine Erkundungsreise*. Berlin: Taschenbuchverlag.

Wunderwerke unserer Epoche 1952. *Die Großbauten der Stalinschen Epoche*. Berlin: Dietz.

*Zarevo nad Volgoj* 1959. Moskva.

# Chapter 14

# The Challenges of Transportation Integration in the U.S.A., 1890–1960

Bruce E. Seely

This chapter will review aspects of the experience of the United States in the process of infrastructure integration, with an awareness of the findings presented by other contributors to this volume. The focus will fall upon transport systems, an emphasis that aligns nicely with the topics examined by most of the other authors; in any case it would be impossible to examine all USA infrastructure networks. From the outset, however, it is necessary to recognise that the continental scale of the USA requires at least one fundamental shift in perspective from that adopted by most other contributors. Rather than exploring how a single infrastructure network was integrated (or not) across national boundaries, the stance adopted for this chapter is to examine the challenges of connecting different American transport sectors into a single functioning network. To be sure, several authors discuss more than one mode of transport, but multi-modal questions are not the central focus of their chapters. This chapter could have reviewed in detail the development of the standards, regulatory structures, legal mechanisms and other activities that promoted operational integration within a single infrastructure sector, for those problems were every bit as challenging in a North American context as in Europe. But the challenges of integration *across* sectors is the essential issue for the United States, for the continental sweep of the nation largely removed the political boundaries that so complicated infrastructure integration in Europe. Thus for the United States, a discussion of intermodal integration offers a more interesting story, even while this topic also adds a different dimension to the conversation among infrastructure within this volume.

The chapter that follows first reviews a handful of themes developed by other contributors, with the goal of applying those themes to the history of transport integration in the United States. Then the focus will shift to a rapid historical survey of efforts to integrate USA transportation sectors into a working network. A concluding section ties the USA experience to those thematic elements identified in the beginning section of the chapter. Patterns and issues that characterise infrastructure integration in Europe and elsewhere can be found in in the United States, but enough variation exists to require careful and nuanced comparisons.

## 1. Thematic Continuities Related to Integration

The other chapters within this volume significantly advance discussions about the question of infrastructure integration, a topic that has not attracted much scholarly attention. This is unsurprising since the term *infrastructure* came into widespread usage only after World War II. Historians of technology began to focus on the subject after Thomas P. Hughes in particular demonstrated the value of exploring large-scale technical systems (Mayntz and Hughes 1988, La Porte 1991, Summerton 1994, also Coutard 1999, and Joerges 1988).

The Tensions of Europe (ToE) project furthered that emphasis, and scholarly activities within that project in turn influenced both the conception of this volume and the content and structure of many of its chapters. The ToE initiative not only identified the challenges and the consequences of integrating essential transport, power, and communication systems across national boundaries, but also suggested that transnational frames of reference may be more appropriate than a national perspective for understanding the development and management of complex infrastructure networks. Specifically, scholars within the ToE group argued that efforts to integrate technical infrastructure systems highlighted the virtues of stronger economic and political connections, thereby providing pivotal justifications for the many proposals to politically integrate Europe after 1945. These infrastructure initiatives stand out much more clearly from a supra-national viewpoint (Tensions of Europe, 2007–2011). Such a stance also is consistent with the recent desire of many historical and social sciences scholars to supplement the nation-state with global or world-historical perspectives as framing devices. Moreover, this viewpoint also seems to be applicable outside of Europe, as the chapters on Africa and Asia demonstrate.

Appropriately, several of this book's chapters offer methodological or conceptual approaches to understanding infrastructure integration. Thus Gerold Ambrosius (2013) seeks to identify archetypes that can be applied to all infrastructure systems. Significantly, he finds that questions of standards – both in terms of technical interconnection and operating processes – provide a useful point of departure for most conversations about integration of systems. Christian Heinrich-Franke's essay (2013b) is perhaps the most explicitly conceptual, as he emphasises the difficulties that face comparisons across sectors and time periods. And S. Waqar H. Zaidi's essay (2013) on functionalism and neo-functionalism highlighted the thinking of British international relations expert David Mitrany for developing general understandings of the integration and management of large infrastructures operating on the international scale. In other words, as technical systems began to cross national borders, system needs forced innovative new approaches to structure, institutions and governance.

Most of the other chapters in this voume are consistent with the conceptual frameworks proposed by these three authors. Perhaps most significantly, almost every chapter suggests that the most difficult challenges in developing and operating large infrastructure systems were not, in the end, technical in nature.

Rather, the truly complicated obstacles to integration were primarily political. This does not mean that integration proponents did not encounter difficult-to-resolve technical obstacles. Yet in chapter after chapter, the political challenges were more resistant to solution because the existence of national boundaries, cultures and political and economic systems complicated the extension or seamless operation of infrastructure networks across Europe. Martin Schiefelbusch's (2013) account of railroad freight transport, for example, indicates the tension between national interests and transnational structures for operating the freight railroad network long remained a pivotal factor. Similarly, Zaidi's (2013) account of the wartime effort by British entrepreneur George de Lengerke to create a transnational airline for Europe shows the range of domestic and international political issues upon which integration discussions turned. Several chapters recognise the enormous influence of the United States, both technically and politically, in post-1945 infrastructure programs, a situation that led many European nations to assert national or regional independence and control over key infrastructures.

Tomasz Komornicki's account (2013) of the changing position of Poland's road and rail networks within the larger systems of Europe also highlights the explicitly political dimensions of infrastructure integration. For several hundred years, Poland's geographic position between Europe and Russia caused its transport networks to be shaped by the dominant powers: first Germany, then the Soviet Union, and more recently the European Union. Thus the Soviet Union re-oriented Poland's traditional transport links from Europe and Germany after 1947 to point to the USSR and Ukraine. Ties to the West reopened slowly after 1970, but the demise of the Soviet Union after 1990 prompted another realignment of Poland's road and rail systems toward Germany. The geopolitical considerations, as well as national and local politics, displayed in Poland also affected the outcome of integration projects outside of Europe as well. The account of Kenya's rail and highway developments by Jacqueline Klopp and George Matajuma (2013) highlights not only the initial impact of colonial decisions by British authorities, settlers, and economic interests, but also the long term impact of internal and regional politics in the period after Kenyan independence. Irene Anastasiadou's summary (2013) of the effort to develop a transnational railroad following the ancient central Asian silk road captures a similar colonial and postcolonial situation. It is fascinating that current economic considerations and the global movement of containers may prove sufficiently persuasive to overcome national concerns that have so far prevented construction of the necessary interconnections.

Another central continuity in infrastructure systems concerns the key role of engineers and technical experts in promoting integration plans. Mitrany's conception of functionalism offers an explanation for the authority gained by engineers, with his description of how the complexity of the technical and political issues surrounding transnational systems produced the conditions that gave engineers more leverage (Zaidi 2013). Engineers not only drove technical discussions, but often were the only actors to wield sufficient influence and authority in the political realm to push forward transnational projects. Indeed, some networks seemed so inherently technical (e.g., satellite communication) that engineers usually dominated the conversations. Seen from another perspective, Ambrosius (2013) argues that the essential importance of standards at the transnational level drew engineering experts to the forefront of even political decision-making. Vincent Lagendijk's discussion (2013) of electricity networks in Europe, painted a similar picture.

A final thread woven through the chapters of this volume concerns the relatively few cases of successful infrastructure integration. Indeed, the weight of the evidence shows sub-marginal accomplishments or overblown and unrealistic expectations., suggesting that failure is a common outcome of infrastructure projects. Jens Ivo Engels (2013) in particular discusses the limits of infrastructure as a unifying activity, arguing that fragmentation and segregation are highly likely to result from the development of infrastructure systems. He argues transnational political ties are simply weaker than national connections, with the result that trans-national projects rarely arouse deep passion and excitement, except perhaps among technical experts. Several other chapters reinforce Engels' argument, demonstrating the continued inability to establish an integrated railroad freight tariff within Europe (Heinrich-Franke 2013a); the failure of Kenya to integrate roads and rails despite compelling needs (Klopp and Matajuma 2013); the failed airline plan for post-World War II Europe and Britain (Zaidi 2013); or the apparent inability to launch the new silk road railroad across central Asia (Anastasiadou 2013). Martin Schiefelbusch (2013) found that plans for supra-national railroad operations never have been fully accepted in Europe, despite the successful introduction of high-speed intercity trains, so that the European rail network remains 'a simple juxtaposition of national systems' rather than the integrated supra-national system envisioned by some advocates.

Thus three overarching themes emerge from the various chapters of the volume: the basically political nature of the difficult problems of system integration, the role of engineers and technical expertise in promoting and implementing integrated systems, and the tendency of many integration plans to fail or not reach all of their goals. These lessons are reinforced by the transportation integration experiences of the United States.

## 2. Transportation Integration in the United States[1]

### *2.1 Integration during the Nineteenth Century*

Comparisons of transport infrastructure integration in Europe and elsewhere with the United States must begin from an understanding of several basic differences. These differences affected efforts to integrate *within* sectors, but had special

---

1   This discussion of U.S. transportation policy will draw extensively from the author's previously published work (Rose, Seely and Barrett 2006; Seely 1987).

significance for efforts to integrate *across* sectors. Of primary importance is the size and continental scale of the United States, for the 3,000-mile wide land mass between the Atlantic and the Pacific oceans, as well as the geographic variation within that territory, affected the planning and construction of every national transport system. Amidst this geographic variation, however, the political control of this territory was not subdivided among the states so that border crossings were removed as obstacles that could restrict transport, especially after the Supreme Court ruled that state governments could not regulate or control interstate commerce (Taylor 1951: 58–9).

During the nineteenth century, leaders of the United States linked development of transportation systems to national goals, considering the movement of people and information essential for both the economy and the functioning of a representative democracy. Moreover, after 1790 political leaders placed emphasis upon transportation to support territorial expansion. Thus significant public support existed for the development of transport systems. National leaders enthusiastically embraced one transport technology after another as the solution to the nation's transport problems, beginning with coastal shipping, progressing to the construction of roads and turnpikes, and then turning to steamboats on larger rivers. Canals were favored during the 1820s and 1830s, before the railroads emerged as a transportation technology capable of serving the entire nation. During their different moments in the sun, each mode of transport generated significant public enthusiasm by promising faster, more reliable and less expensive means of moving people and goods. Government officials at every level supported and subsidised privately constructed and operated transport networks. Canals in particular benefitted from this mindset, but every form of transport received public assistance (Dunbar 1915; Durrenberger 1968; Fishlow 1972; Goodrich 1960; Miner 2010; Paskoff 2007; Taylor 1951; Woods 1997).

These political and technical patterns resulted in a rapidly changing and chaotic transportation situation. Importantly, this transport environment was characterised more by competition than collaboration. Steamboats competed successfully against roads, but eventually succumbed to railroad service. Similarly canals hurt turnpikes, but railroads eventually put most canals out of business. Indeed, railroads eventually proved dominant over every mode of transport. This competitive structure reflected a public policy regime that exhibited little concern for connections between modes. The integration of transport systems was not a topic of national conversation. Then again, national policy showed little concern for achieving integration *within* modes either. No national system existed for setting standards until the twentieth century, and even then the participation of transport firms was not compulsory (Seely 1984). Railroads were constructed with tracks of many widths (gauges), often forcing freight and passengers connecting to a different line to physically transfer to different equipment. In any event, few rail terminals served more than one line. With the interchange of equipment almost impossible, the transfer of people and freight was a slow, expensive proposition.

After 1860, however, an integrated system of sorts slowly emerged without central guidance simply because railroads became the nation's primary transportation agency. Experiences in the northeast and midwestern sections of the country demonstrated that railroads were suited to service over long distances, sometimes difficult geography, and the changeable seasons that characterised the interior of the country. An unevenly distributed network of 30,000 miles of railroad track was in place at the start of Civil War, and the conflict soon erased any doubts about the value of a functioning rail system on a national scale. The first transcontinental lines, for example, were quickly authorised by a Union government determined to keep California and Oregon connected to the east – and the Union. Once again, indirect federal assistance (land grants) helped offset construction costs, this time for rail lines built before people and traffic were in place. By the 1880s, the railroad network effectively *was* the transport system of the United States, and over the next decades the network grew until total mileage peaked in 1916 at 254,000 miles (Klein 1987; Martin 1992; Stover 1961; Taylor, 1951; White 2011).[2]

The dominance of the railroads actually removed pressure for serious discussions about intermodal transportation integration. Highways, especially farm-to-market roads in rural areas, provided the 'last mile' of service to and from rail stations, but were considered a local, not national, transport element. No state even identified a road system until New Jersey and Massachusetts created networks in the mid 1890s. No national road system existed until 1921 (Seely 1987: Chapter 1). Canals and inland waterways, including steamboats on the Mississippi River, had steadily declined in importance after the 1850s except on the Great Lakes. Despite efforts to promote water transport as an alternative to railroads, waterways were best suited for low value bulk commodities because service was slow and utterly dependent upon water levels and temperatures. Railroads normally had the advantage of speed and reliability, although many lines were fragile until the end of the nineteenth century. Richard White argues even the transcontinental railroads were primarily regional freight carriers freight during the nineteenth century (White 2011: 162–73).

The seamless movement of traffic developed only slowly because of the often-discussed lack of integration among the different carriers. The key issues that had to be addressed included acceptance of a standard railroad track gauge (eventually 4 feet, 8–1/2 inches), determination and installation of standard train brakes and couplers on all cars, the definition of time zones, and the development of 'union stations' that served multiple railroads in a few larger cities. Standards in all of these areas were essential to easier interchange of people and goods between different carriers. They were only grudgingly adopted after 1880, in part because no government agency could compel action by rail carriers. Instead, professional

---

2  Richard White (2011) offers a revisionist interpretation of the transcontinental railroads that suggests that during the nineteenth century rail carriers provided a national transport system in only a very limited fashion.

railroad technical organisations, the private rail carriers, and equipment suppliers such as Westinghouse Air Brake shaped the outcome of discussions about operational and equipment standards. In the end, railroads adopted industry-wide standards only when they chose to do so. The Pennsylvania Railroad, for example, used research and testing in its shops to implement product specifications after 1880, but gaining acceptance for industry-wide standards among multiple corporations was much more difficult. The emergence of professional organisations of railroad engineers helped somewhat by disseminating information. Even so, a standard gauge and national time zones were accepted only when interchange of equipment or the scale of operations reached a certain threshold in the 1880s (Puffert 2009; Stover 1993; Taylor and Neu 1956). And railroad officials hesitated much longer before installing expensive safety appliances (couplers and brakes). They waited until superior, workable, and affordable solutions were available, and until they had confidence all carriers would install compatible devices. Public pressure generally was not sufficient to speed up decisions by private carriers on these standards (Usselman, 2002).

## 2.2 Integration during the Twentieth Century

Although not immediately apparent to most observers, the transportation dominance of railroads began to erode after 1910 as multiple new technologies began to deliver competing transportation services. The revival of interest in inland waterways began before 1900 brought significant federal investments in river improvements such as dams and locks, while after 1920 civil aviators began to carry mail and a few daring passengers long distances. But the most important new option was the highway-motor vehicle system, as cars along with busses and trucks offered viable alternatives to the rail carriers. The challenge facing both transportation providers and policy makers was how to mesh these new transport technologies into a functioning transport system. Nearly all participants in the resulting policy debates failed to accomplish such an outcome, in part because public policy in the USA regarding transportation remained largely unchanged from the nineteenth century. Thus government assistance and subsidies continued to encourage transportation development under the aegis of competitive private corporations operating a series of independent networks.[3] Indeed public policy during the twentieth century not only favoured *competition* between technologies, but often discouraged *cooperation* between the different systems (Miller 1933; Rose, Seely and Barrett 2006: 33–49). In sum, the United States did not develop a genuinely integrated transportation system.

This policy framework rested in part upon deep-rooted and passionate reactions to the enormous economic and political power that rail carriers had

---

3   The acceptance by local, state, and federal governments of responsibility for a highway system was the most obvious exception to this pattern, a point that promoted angry complaints from many rail carriers.

exercised during the nineteenth century. Many citizens and government officials alike viewed railroads as dangerous economic predators, a stance that had helped justify creation of the Interstate Commerce Commission (ICC) in 1887 to regulate American railroads. Many assumed that railroad rates deliberately discriminated against farmers and western merchants, prompting strident calls for state and federal regulation to prevent such abuses. Later, regulatory restrictions on rail carriers came to be seen as mechanisms for protecting new technological competitors that could eventually grow into systems capable of breaking the monopoly power of railroads corporations. To be sure, the mix of reasons justifying the regulatory regime for railroads was more complex than simply punishing railroads for bad behavior (Berk 1994; Deverall 1994; Kerr 1968; Klein 1900; Kolko 1965; White 2011; Wiebe 1967). But whatever the motivations for government regulation, the foundation of railroad regulatory policy was a mode-by-mode view that treated each transportation technology as an independent entity (Meyer 1959). And after 1900, a more aggressive approach to this end resulted in actions to wall off railroads from all other modes of transport. Furthermore, the policies applied to the different modes were deeply inconsistent, with railroads tightly controlled while their competitors were encouraged and even subsidised.[4]

Congress established this approach by requiring the ICC to enforce increasingly restrictive regulatory rules applied to railroads. Initially, shippers were forced to challenge railroad rates in the courts. The futility of this approach, in an era when railroad corporations might 'control' judges, prompted Congress to tip the regulatory scales against the railroads by moving the appeal process to the ICC. Congress also allowed the ICC to shape railroad business practices, as when the commission established accounting standards and reporting practices. Then after 1905 railroad executives were required to secure ICC permission *before* raising rates. This action effectively froze railroad rates for more than a decade. The ICC also launched a long-running study of the valuation of railroad property in order to determine whether railroads had issued watered stock and in other ways misled investors. Finally in 1912, Congress prohibited any railroad-owned ship from transiting the Panama Canal in order to prevent rail carriers from branching into steamship operations (Berk 2004; Bryant 1988; Ely 2001; Miranti 1989).

Even as federal regulation began to restrict the activities of railroads, other modes of transport encountered national and state policies designed to stimulate and encourage their development. Transport on inland waterways, for example was seen by many in the Midwest as an ideal alternative to railroads for the shipment of grain. The result was a federal construction program for locks and dams that

---

4   This approach to railroads resembled the public policy strategy most cities adopted toward street railways after the mid 1920s – treat street car franchises as corrupt and unresponsive organisations while catering to the modern and democratic automobile. Urban accommodations to motorists included the development of boulevards, the construction of parking lots, the implementation of traffic-moving strategies such as one-way streets and signals, and so on (Barrett 1983).

between 1890 and 1931 expended more than $1.3 billion. By 1920, the states had spent another $300 million (Baker 1920; Dearing and Owen 1949: 81–104; Hull and Hull 1967; Howe 1969; Miller 1933).[5] In addition, during World War I Congress chartered the Inland Waterways Corporation to operate a barge line on several southern rivers specifically to create competition with rail carriers. Federal subsidies alone allowed the company to survive (Baker 1920: 85–9; Dimock 1935; N.A. 1928).

Civil aviation was even more dependent upon government supports and a favourable regulatory environment. Early federal assistance included navigation beacons and other safety mechanisms, as well as research conducted by the National Advisory Committee on Aeronautics to assist airplane manufacturers. Local governments often constructed airfields. But flying remained more about exotic excitement than functional transport until the introduction of air mail subsidies in 1925 provided the economic foundation for passenger service. By allowing carriers a profit carrying the mail, passenger service could be added at very little risk to the business (Kommons 1978, Lee, 1984; van der Linden 2002, Smith 1965, Whitnah 1966).

Railroad executives bitterly protested that property taxes on their rights-of-way allowed local, state, and federal governments to subsidise (unfairly, they complained) highway construction projects. The federal highway program began in 1893, as railroad corporations ironically with railroad supported an improved network of farm-to-market roads allowing rural residents better access to railroad stations. Bicyclists also supported good roads. Indeed, only after 1905 were automobile makers strong proponents of improved roads. In 1912, Congress funded a trial program to improve roads for rural mail delivery, and in 1916 established a permanent post-road construction program with a modest appropriation of $5 million that grew to $25 million in 1921. But the war demonstrated the potential of motor transport and in 1921 Congress authorised a $75 million a year for network of primary and secondary highways linking major cities.

Those railroad executives had reason to be concerned. As early as 1915, a few large railroads experienced noticeable losses of short-haul freight traffic and reduced numbers of passengers (N.A. 1920; Moulton 1933: 530–46, Miller 1933: 571–601, Dearing 1941, Seely 1987). By 1930, automobiles accounted for 80 to 90 per cent of *all* intercity passenger-miles, an amazingly rapid change. Rail carriers lost 20 billion passenger-miles and the reductions were catastrophic at some companies. The Cotton Belt Railway, for example, delivered only 29 million passenger-miles in 1929, compared to 137 million passenger-miles in 1920, while executives at the Missouri Pacific witnessed a decline in passenger revenue from $21 million in 1920 to $10 million in 1930. During the interwar period, trains continued to carry long-distance passengers, but automobiles also challenged that service (N.A. December 1930,

---

5 Ironically, even at the time is was not clear that these investments made economic sense (Moulton, 1912).

Miller 1933: 582, 598–9; Moulton 1933: 517–46). Clearly, the railroad monopoly on transportation had ended.

Given this changing situation, railroad officials protested even more vehemently about the differences in the regulatory treatment of their competitors. While rail carriers were tightly controlled by the ICC, early air lines first faced few safety or operating standards. Bus and truck operators initially encountered few rules, and then after state regulation of trucks and busses was forbidden in 1926, even fewer restrictions applied to busses and trucks until federal regulations were took effect in 1935. Worse in the eyes of railroad executives, after the mid 1920s state officials closely scrutinised railroad involvement in bus and motor trucking companies (Childs 1981: 93–5; Walsh 2000: 95–8, 138–403; also N.A. April 1926, N.A. June 26, 1926, N.A. 29 January 1928). In effect, railroads only rarely could directly operate (and thus control, as anti-railroad groups feared) new transport technologies. This proved particularly frustrating to the largest rail carriers, a few of whose executives sought to improve profitability and provide better service with alternative technologies. These rail carriers, especially the Pennsylvania, saw an opportunity for genuine integration and aggressively explored using trucks and busses to replace trains on lightly used branch lines. Forced by regulatory decisions to establish wholly-owned subsidiaries that could not be directly integrated with rail management, the number of railroads with motor bus operations rose from ten in 1925 to 78 in 1929. The Pennsylvania and the New York Central Railroads, among others, also explored using freight containers that could be shipped by rail and delivered to a final destination on a truck (Emery 4 January 1930; Seely 1986; Walsh 2000: 81–2, 92–5). The most imaginative – or perhaps quixotic – program was combined train and air service between New York City and Los Angeles launched in 1929 by the Pennsylvania and Santa Fe Railroads and TAT airlines. Intrepid travellers rode a night sleeper car from New York's Pennsylvania Station to Columbus, Ohio, where they boarded a plane and flew to Indianapolis, St. Louis, Kansas City, Wichita and Waynoka, OK. Passengers changed to the Santa Fe overnight train to Clovis, New Mexico, before catching a plane for the final leg into Los Angeles. Total travel time was 48 hours. In 1929, W.W. Atterbury, president of the Pennsylvania, captured the hopes of some advocates of integrated transportation with the statement, 'We are no longer railroads alone; we are transportation companies' (N.A. 6 July 1929, also N.A. 6 April 1929, N.A. 20 July 1929).

Atterbury's views about integration resonated with ideas then percolating among transportation industry leaders and academic economists such as Emory Johnson at the Wharton School and the U.S. Chamber of Commerce. He and colleagues advocated co-ordination instead of wasteful competition, consistent regulation of all forms of transportation, and acceptance of railroad use of motor vehicles to replace trains in certain situations (N.A. 1923). The onset of the Depression prompted even greater interest, as G. Lloyd Wilson, another Wharton School economist, provided the operative definition of the term *transportation co-ordination* when he proposed that 'each unit occupies its proper place', as determined by some measure of relative economic efficiency (Wilson 1930: 2–3, also Miller 1933). Three additional reports between 1932 and 1934 echoed this sentiment. Emory Johnson (1938: 645) prepared a second report urging that rail carriers be permitted to engage in water and highway transport with all transport regulated under the same federal policy. William J. Cunningham (1934), the Harvard Business School's James J. Hill Professor of Transportation, in a report prepared for the National Highway Users Conference, and Harold G. Moulton, in a book for the Brookings Institution which he headed, endorsed similar positions. Moulton concluded American transportation agencies worked 'at cross purposes ... Instead of a unified program of regulation designed to promote a common objective, we have a series of unrelated and often antagonistic policies carried out by a variety of government agencies'. His solution matched the other reports – let the transport mode best suited in terms of cost do the job by ending government subsidies and unequal regulation (Moulton 1933: 881, also N.A. 18 February 1933, N.A. 18 March 1933).

The near-collapse of the nation's railroads in 1932–33 prompted efforts to actually act upon these ideas. President Franklin Roosevelt appointed Interstate Commerce Commission Joseph Eastman as Federal Co-ordinator of Transportation. Eastman echoed the academic economists, arguing for a co-ordinated system, not separate technologies embedded in individual industries. 'The ideal to be achieved', he wrote, is to 'utilize each agency in the field for which it is best fitted and discourage its use where it is uneconomical or inefficient' (Eastman 30 September 1933). In supporting such steps as co-ordinated rail/truck at rail terminal operations, Eastman argued, 'this work is by no means a mere attempt to bolster up the railroads at the expense of the trucks', but an attempt to develop better transportation' (N.A. 28 October 1933). And in May 1934, he stated wasteful competition between railroads, trucks, and barges had to end; coordination would 'make the national railroad system more efficient and economical' (N.A. 17 March 1934).

After two years of studies, Eastman proposed legislation to assign regulatory control of all modes of transportation to the ICC and remove subsidies that skewed the economic value of each transport technology. After much debate, Congress rejected Eastman's program in 1935 because of opposition from the motor transport industries, civil aviation, and waterways. All preferred the older pattern of restrictive regulation of the railroads. In the end Congress accepted Eastman's plan for ICC regulation of trucks and busses, but under separate divisions that operated with almost complete autonomy. This Motor Carrier Act of 1935 directed ICC regulators to 'recognize and preserve the inherent advantages of each' mode of transportation (Rose, Seely, and Barrett 2006: 71). In other words, trucks and busses were still shielded from the railroads. And aviation had sought similar special treatment, arguing vehemently that they could not be grouped with the railroads and motor vehicles. Aviators argued, in the words of one supporter that 'a more intimate touch and knowledge of the needs' of aviation was required (Rhyne 1939: vi).

The passage of the Transportation Act of 1940 captured the true state of affairs in terms of transportation co-ordination and integration. The bill opened with the statement that it was national policy to regulate all forms of transportation 'to the end of developing, coordinating, and preserving a national transportation system by water, highway, and rail …'. Therefore, responsibility for regulation of inland waterways was added to the ICC's responsibilities. But Congress also ordered that the advantages of water shipment be preserved. Thus as had happened with truck and bus regulation, a single agency was in charge, but the policies for each mode of transportation were wildly different. Thus U.S. national transportation policy remained fragmented, with different regulatory regimes for different modes preventing anything like integrated transport service (Rose, Seely, and Barrett 2006: Chapter 2).

The continued failure to achieve integration was perhaps most apparent after World War II in the regulatory treatment of railroad attempts to utilise motor vehicles. While airlines began to offer large-scale service with the assistance of a variety of subsidies, and as the highway system culminated in the very expensive Interstate Highway Program of the 1950s, rail carriers struggled to remain viable under rigid and restrictive regulations. Railroad attempts to acquire bus or truck companies had encountered resistance during the 1920s and 1930s, but by creating subsidiary companies, many rail carriers developed some truck and bus operations. But after the war, according to transportation economist Emory Johnson, federal legislation kept 'railroad participation in motor transportation, and the co-ordination of rail and motor transportation, within narrow limits' (Johnson 1947: 125–6, also N.A. 8 October 1938a, N.A. 8 October 1938b). This regulatory policy continued well into the post-war period, as evidenced by the ICC's effort to prevent the development of 'piggyback' service (railroad shipment of truck trailers) during the 1950s. The introduction of ocean shipping of truck-sized freight containers also encountered regulatory obstacles, which Malcolm MacLean, a key innovator in this area, resolved only by scheduling his ships to call at Havana in order to escape U.S. regulatory jurisdiction (Cudahy 2006; DeBoer 1992; Donovan 2000; Donovan and Bonney 2006; Levinson 2006; Seely 2008).

At the same time, growing dissatisfaction was heard from more and more economists and from elected officials concerning the difficulties and costs imposed by the nation's mode-by-mode regulatory strategy (Meyer 1959). Every U.S. president from Dwight Eisenhower onward grappled with the issue, which has usually been presented using the term *de-regulation*. In fact the policy discussions really emphasised redefining the relationship of transportation modes to each other and the importance of transport to the economy. President Lyndon Johnson attempted to promote transportation integration by creating a new cabinet-level U.S. Department of Transportation (U.S. DOT) to encompass all transportation agencies except the ICC and bring together the dispersed authority related to transport. The new agency did little, however, to integrate the various transportation sectors, as both highway program – U.S. DOT's largest element – and smaller agencies covering aviation, waterways, and transit operated as largely independent offices. The regulatory pattern was untouched and untouchable, since the Interstate Commerce Commission was legally independent of direct day-to-day control by the Executive branch.

Presidents Richard Nixon, Gerald Ford, Jimmy Carter and Ronald Reagan followed with their own attempts to address obvious inefficiencies in transport services, but Nixon backed off because of the politics of entrenched interests. Ford launched more prolonged discussions, while Carter actually pressed legislation through Congress. Finally, in 1980 the integration of modes became possible, first through the de-regulation of the airlines, a move debated during the Ford and Carter administrations and carried out by the Reagan presidency. The end of much of the ICC's regulation of trucking and railroad rates and operational details followed, giving both sectors much freer hands in providing transport services. A key side result was a much greater ability to build cooperative and even collaborative ventures, such as allowing interconnections between trucking and railroads (Rose, Seely and Barrett 2006: chapters 4–9). Reflective of the policy shift was the 1991 highway funding bill, the Intermodal Surface Transportation Efficiency Act, which for the first time highlighted the interaction of the different transportation sectors or modes as the aim of federal funding. Actual adjustments in actions have come slowly over the past two decades, since the modal orientation of national policy is deeply entrenched in the federal agencies, in the Congressional committee structure, and among transportation providers. Understanding how to make intermodal transport work has requires time to learn new possibilities.

## 3. U.S. Transport Integration within the Larger Patterns of this Volume

The three key patterns seen in this volume's chapters about Europe and elsewhere are to a significant extent visible as well in the U.S. experience. First, even this rapid and often undetailed review of the development of transportation policies in the USA strikingly confirms the suggestion that infrastructural integration is largely a political exercise. This political tendency is equally central to both intra-sectoral integration as described in most of the other chapters in this volume or to intermodal concerns as examined in the United States. The key venues for the development and establishment of policy in the U.S. were the Congress and increasingly through the twentieth century, the office of the President. That is, key decisions were made at the very center of national politics in the USA. Moreover, other actors in the drama – workers, shippers, transportation companies, and many other stakeholders – all considered the determination of transportation policy to be a political matter.

This outcome impinges upon the second pattern, which was the role of experts and engineers in the shaping of policy. There is certainly ample evidence that experts were actors in the process of developing and implementing transport policy in many different national settings. Engineers often have assumed leadership roles in activities that extend well beyond than the actual construction of the networks themselves. In the case of U.S. highway policy, that role was substantial, for federal highway engineers served as key arbiters of almost all aspects of policy from the 1890s through the 1960s (Seely 1987). Economists also played a very important role in delineating the U.S. transportation situation, often in ways that helped framed the technical and political debates. The names of many of the leaders in transportation economics and policy have been identified here, yet it is intriguing how often the ideas of economists such as Emory Johnson failed to find their way into policy. In some instances, their recommendations were enacted, but even being written into legislation could not guarantee that their ideas could shape transport as the economists had planned, even when written into legislation (Rose, Seely and Barrett 2006: chapters 1 and 2).

These failures of infrastructure integration efforts further connect the U.S. transportation experience to the pattern found in other chapters in this volume. The most obvious evidence of conformity to the pattern is the fact that the U.S. only recently adopted national policies which place emphasis upon the integration of different transport systems. For most of the nation's history, the exact opposite was the case. Even when a policy statement claimed the goal of advancing connections between different technological systems, in reality the American emphasis upon competition usually resulted in policy prescriptions considering only one mode at a time. Little attention was given to interactions or connections to other modes of transportation. To be sure, attempts were made to accomplish some degree of transportation integration or coordination – many more than are reported in this text. Yet almost all have failed to realise the goals and hopes of their supporters and advocates.

Only in the past 20 years have successful intermodal ports, airports, and rail terminals appeared, driven largely by the adoption of container services, air freight, and other elements of global economic trade. Importantly, many of these innovations came in spite of, not because of, national public policy. Indeed, the best examples of integrated transport operations in the U.S. today may be UPS and FedEx, freight shippers that style themselves logistics firms. Significantly, these companies are the first that can move packages based upon service considerations and cost rather than the particular transport technology they operate.

Thus the transport infrastructure situation in the USA in the early twenty-first century is finally reflecting some of the goals of integrating the different technological systems that have circulated for 90 years or more. The policy environment developed since 1980 allows modes to collaborate rather than simply compete in inefficient ways. The stated goals of U.S. transport policy now stress intermodal operations. And the emergence of UPS and FedEx shows the possibilities of this way of connecting the various transport technologies. These two firms deliver basic transport services, yet they are not recognised as transportation firms. It is interesting to speculate what the transport scene in the USA might look like today like if the regulatory and political climate of the 1920s had allowed American railroads to consider becoming transportation companies and not just railroads.

# References

Ambrosius, G. 2014. Archetypes of International Infrastructural, in *Linking Networks: The Formation of Common Standards and Visions for Infrastructure Development*. Martin Schiefelbusch and Hans-Liudger Dienel, eds. Farnham: Ashgate, 35–50.

Anastasiadou, I. and Tympas, A. 2014. Iron Silk Roads: Comparing Interwar and Post-war Transnational Asian Railway Projects, in *Linking Networks: The Formation of Common Standards and Visions for Infrastructure Development*. Martin Schiefelbusch and Hans-Liudger Dienel, eds. Farnham: Ashgate, 169–86.

Baker, C.W. 1920. What is the Future of Inland Water Transportation? *Engineering News-Record* 84(1 January 1920): 19–29; (8 January 1920): 85–9; (15 January 1920): 137–44; (22 January 1920): 184–91; (29 January 1920): 234–42.

Barrett, P. 1983. *The Automobile and Urban Transit: The Formation of Public Policy in Chicago, 1900–1930*. Philadelphia: Temple University Press.

Berk, G. 1994. *Alternative Tracks: The Constitution of American Industrial Order, 1865–1917*. Baltimore: The Johns Hopkins University Press.

Berk, G. 2004. Whose Hubris? Brandeis, Scientific Management, and the Railroads, in *Constructing Corporate America: History, Politics, Culture*. K. Lipartito. and D. Sicilia, eds. Oxford: Oxford University Press, 120–48.

Bryant, K., 1988. *Railroads in the Twentieth Century; Encyclopedia of American Business History and Biography*. New York: Bruccoli Clark Layman/Facts on File.

Childs, W.R. 1981. *Trucking and the Public Interest: The Emergence of Federal Regulation, 1914–1940*. Knoxville: University of Tennessee Press.

Coutard, O. 1999. *The Governance of Large Technical Systems*. London and New York: Routledge.

Cudahy, B.J. 2006. *Box Boats: How Container Ships Changed the World*. New York: Fordham University Press.

Cunningham, W.J. March 1934. The Correlation of Rail and Highway Transportation. *American Economic Review* 24(1): 48.

Dearing, C. 1941. *American Highway Policy*. Washington, DC: The Brookings Institution, 1941.

Dearing, C., and Owen, W. 1949. *National Transportation Policy*. Washington, DC: Brookings Institution.

DeBoer, D.J. 1992. *Piggyback and Containers: A History of Rail Intermodal on America's Steel Highway*. San Marino, CA: Golden West Books.

Deverell, W. 1994. *Railroad Crossing: Californians and the Railroad, 1850–1910*. Berkeley, CA: University of California Press.

Dimock, M.E. 1935. *Developing America's Waterways; Administration of the Inland Waterways Corporation*. Chicago, IL: University of Chicago Press.

Donovan, A. 2000. Intermodal Transportation in Historical Perspective. *Transportation Law Journal* 27(3): 317–44.

Donovan, A. and Bonney, J. 2006. *The Box That Changed the World: Fifty Years of Container Shipping – An Illustrated History.* East Windsor, NJ: Commonwealth Business Media.

Dunbar, S. 1915. *A History Of Travel In America.* 4 vols. Indianapolis: Bobbs-Merrill.

Durrenberger, J. 1931, 1968. *Turnpikes: A Study Of The Toll Roads Movement in the Middle Atlantic States and Maryland.* Cos Cob, CT: Jon E. Edwards.

Eastman, J.B. 30 September 1933. The Work of the Federal Co-ordinator. *Railway Age* 95: 468–71.

Ely, J. 2001. *Railroads and American Law.* Lawrence: University Press of Kansas.

Emery, J.C. 4 January 1930. Motor Transport Looms Larger in Railway Picture. *Railway Age* 88: 106–7.

Engles, J.I. 2014. Infrastructure and Fragmentation: The Limits of the Integration Paradigm, in *Linking Networks: The Formation of Common Standards and Visions for Infrastructure Development.* Martin Schiefelbusch and Hans-Liudger Dienel, eds. Farnham: Ashgate, 19–34.

Fishlow, A. 1972. 'Internal Transportation', in *American Economic Growth: An Economist's History of the United States.* Lance Davis, et al. New York: Harper & Row, 468–574.

Goodrich, C. 1960. *Government Promotion of American Canals and Railroads.* New York: Columbia University Press.

Heinrich-Franke, C. 2014a. Functionalistic Spill-over and Infrastructural Integration: The Telecommunication Sector, in *Linking Networks: The Formation of Common Standards and Visions for Infrastructure Development.* Martin Schiefelbusch and Hans-Liudger Dienel, eds. Farnham: Ashgate, 95–114.

Heinrich-Franke, C. 2014b. Methodological and Conceptual Challenges of Comparisons across Sectors and Periods, in *Linking Networks: The Formation of Common Standards and Visions for Infrastructure Development.* Martin Schiefelbusch and Hans-Liudger Dienel, eds. Farnham: Ashgate, 9–18.

Howe, C., et al. 1969. *Inland Waterway Transportation: Studies in Public and Private Management and Investment Decisions.* Baltimore: Resources for the Future, Johns Hopkins University Press.

Hull, W.J., and Hull, R.W. 1967 *The Origin and Development of the Waterways Policy of the United States.* Washington: National Waterways Conference, Inc.

Joerges, B. 1988. *Large Technical Systems: The Concept and the Issues.* Berlin: Wissenschaftszentrum Berlin für Sozialforschung.

Johnson, E. 1938. *Government Regulation of Transportation.* New York: D. Appleton-Century Company, Inc.

Kerr, K., 1968. *American Railroad Politics, 1914–1920: Rates, Wages, and Efficiency.* Pittsburgh: University of Pittsburgh Press.

Klein, M. 1987. *Union Pacific: Birth of a Railroad, 1862–1893.* vol 1. *Union Pacific: The Rebirth, 1894–1969.* vol. 2. Garden City, NY : Doubleday.

Klein, M., 1990. Competition and Regulation: The Railroad Model. *Business History Review* 64 (2): 311–25.

Klopp, J. and Makaluma, G. 2014. Transportation Infrastructure in East Africa in a Historical Context, in *Linking Networks: The Formation of Common Standards and Visions for Infrastructure Development*. Martin Schiefelbusch and Hans-Liudger Dienel, eds. Farnham: Ashgate, 115–36.

Kolko, G., 1965. *Railroads and Regulation, 1877–1916*. Princeton, NJ: Princeton University Press.

Kommons, N.A. 1978. *Bonfires to Beacons: Federal Civil Aviation Policy under the Air Commerce Act, 1926–1938*. Washington, DC: U.S. Department of Transportation.

Komornicki, T. 2014. Spatial and Social Effects of Infrastructural Integration in the Case of Polish Borders, in *Linking Networks: The Formation of Common Standards and Visions for Infrastructure Development*. Martin Schiefelbusch and Hans-Liudger Dienel, eds. Farnham: Ashgate, 187–207.

Lagendijk, V. 2014. From Liberalism to Liberalisation: International Electricity Governance in the Twentieth Century, in *Linking Networks: The Formation of Common Standards and Visions for Infrastructure Development*. Martin Schiefelbusch and Hans-Liudger Dienel, eds. Farnham: Ashgate, 137–50.

La Porte, T.R. 1991. *Social Responses to Large Technical Systems: Control or Anticipation*. Dordrecht: Kluwer.

Lee, D.D. 1984. Herbert Hoover and the Development of Commercial Aviation. *Business History Review* 58 (1): 78–102.

Levinson, M. *The Box: How the Shipping Container Made the World Smaller and World Economy Bigger.* Princeton, NJ: Princeton University Press.

Martin, A. 1992. *Railroads Triumphant: The Growth, Rejection, and Rebirth of a Vital American Force*. New York: Oxford University Press.

Mayntz, R., and Hughes, T.P. 1988. *The Development of Large Technical Systems*. Frankfurt am Main, Campus Verlag; Boulder, CO: Westview Press.

Meyer. J.R., et al. 1959. *The Economics of Competition in the Transportation Industry*. Cambridge, MA: Harvard University Press.

Miller, Sidney, 1933. *Inland Transportation; Principles and Policies*. New York: McGraw-Hill Book Company, Inc.

Miner, C. 2010. *A Most Magnificent Machine: America Adopts the Railroad, 1802–1865*. Lawrence: University of Kansas Press.

Miranti, P. Autumn 1989. The Mind's Eye of Reform: The ICC's Bureau of Statistics and A Vision of Regulation, 1887–1940. *Business History Review* 63: 496–9.

Moulton, H.G. 1912. *Waterways versus Railways*. Boston: Houghton Mifflin Company.

Moulton, H.G. 1933. *The American Transportation Problem*. Washington, DC: The Brookings Institution.

N.A. 17 January 1917. Motor-Car Travel Greater than Railroad Travel *Literary Digest* 54: 164, 166–7.

N.A. 1923. *Relation of Highways and Motor Transport to Other Transportation Agencies*. Washington, DC: U.S. Chamber of Commerce.

N.A. April 1926. Hearings on Bill to Regulate Motor Bus Traffic Started. *Bus Transportation* 5: 195–201.

N.A. 26 June 1926. State Commissioner Comment on Highway Transport. *Railway Age* 80: 1977–78.

N.A. 29 January 1928. Examiner's Report on Motor Transport. *Railway Age* 84: 269–77.

N.A. 24 October 1928 and 13 November 1928. What the Barge Lines Are Doing. *Railway Age*. 85: 801–4, 865–9.

N.A. 6 April 1929. Should Railroads Engage in All Forms of Transportation? *Railway Age* 86: 769.

N.A. 6 July 1929. The T.A.T. Air Rail Service. *Railway Age* 87: 12–16.

N.A. 20 July 1929. Railroads Take to the Air. *Literary Digest* 102: 8.

N.A. December 1930. Bus and Rail Men Testify Before I.C.C. *Bus Transportation* 9: 661.

N.A. 18 February 1933. National Transportation Committee Makes Comprehensive Report *Railway Age* 94: 247–50, 255.

N.A. 18 March 1933. Transport Problem Analyzed By N.T.C. Staff. *Railway Age* 94: 399–401.

N.A. 28 October 1933. Eastman's Views on Motor Transport Regulation. *Railway Age* 95: 636–7.

N.A. 17 March 1934. Regulation of Water and Motor Carriers Recommended. *Railway Age* 96: 377–85.

N.A. 8 October 1938. Extension of Rail-Motor Co-ordination Vetoed. *Railway Age* 105: 525–6.

N.A. 8 October 1938. No Rail-Truck Tie-up; ICC Refuses Union Pacific, Burlington, and North Western Railroads. *Business Week*: 37–8.

Paskoff, P. 2007. *Troubled Waters: Steamboat Disasters, River Improvements and American Public Policy, 1821–1860*. Baton Rouge, LA: Louisiana State University Press.

Puffert, D. 2009. *Tracks Across Continents, Paths Through History : The Economic Dynamics of Standardization in Railway Gauge*. Chicago, IL: University of Chicago Press.

Rhyne, C.S. 1939. *The Civil Aeronautics Act Annotated*. Foreword by Pat McCarran. Washington, DC: National Law Book Co.

Rose, M.H., Seely, B.E., and Barrett, P. 2006. *The Best Transportation System in the World: Railroads, Trucks, Airlines, and American Public Policy in the Twentieth Century*. Columbus, OH: The Ohio State University Press.

Schiefelbusch, M. 2014. Visions of Rail Development – The 'Internationality of Railways' Revisited, in *Linking Networks: The Formation of Common Standards and Visions for Infrastructure Development*. Martin Schiefelbusch and Hans-Liudger Dienel, eds. Farnham: Ashgate, 73–94.

Seely, B.E. 1987. *Building the America Highway System: Engineers as Policy Makers*. Philadelphia: Temple University Press.

Seely, B.E. 1984. Engineers and Government-Business Cooperation: Highway Standards and the Bureau of Public Roads, 1900–1940. *Business History Review* 58(1): 51–77.

Seely, B.E. Autumn 1986. Railroads, Good Roads, and Motor Vehicles: Managing Technological Change *Railroad History* No. 155: 35–63.

Seely, B.E. 1987. *Building the America Highway System: Engineers as Policy Makers.* Philadelphia: Temple University Press.

Seely, B.E. 2008. *Public Policy and Freight Transportation, 1920–1960: Setting Boundaries and Framing Technical Possibilities:* Paper to the Annual Meeting of the Society for the History of Technology. Lisbon, Portugal, 13 October 2008.

Smith, H.L. 1965. *Airways: The History of Commercial Aviation in the United States.* NY: Russell & Russell.

Stover, John F. 1961. *American Railroads.* Chicago, IL: University of Chicago Press.

Stover, J. 1993. One Gauge: How Hundreds of Incompatible Railroads became a National System. *American Heritage of Invention & Technology* 8(3): 55–61.

Summerton, Jane, 1994. *Changing Large Technical Systems.* Boulder, CO, Westview Press.

Taylor, G.R. 1951. *The Transportation Revolution, 1815–1860.* New York: Rinehart.

Taylor, G.R., and Neu, I.D. 1956. *The American Railroad Network, 1861–1890.* Cambridge, MA: Harvard University Press.

Tensions of Europe 2007–2011. http://www.tensionsofeurope.eu/www/en/research (website accessed 15 July 15 2012).

Usselman, S. 2002. *Regulating Railroad Innovation: Business, Technology, and Politics in America, 1840–1920.* New York: Cambridge University Press.

van der Linden, F.R. 2002. *Airlines and the Mail: The Post Office and the Birth of the Commercial Aviation Industry.* Lexington: University of Kentucky Press.

Walsh, M. 2000. *Making Connections: The Long-Distance Bus Industry in the USA.* Burlington,VT: Ashgate Publishing Company.

White, R. 2011. *Railroaded: The Transcontinentals and the Making of Modern America.* New York: W.W. Norton & Co.

Whitnah, D. 1966. *Safer Skyway: Federal Control of Aviation, 1926–1966.* Ames: Iowa State University Press.

Wiebe, R., 1967. *The Search for Order, 1877–1920.* New York: Hill and Wang.

Wilson, G.L. 1930. *Coördinated Motor-Rail-Steamship Transportation.* New York: Appleton & Company.

Woods, F. 1997, reprint of 1919. *The Turnpikes of New England.* Pepperell, MA: Branch Line Press.

Zaidi, S. 2014. 'Wings for Peace' versus 'Airopia': Contested Visions of Postwar European Aviation in World War Two Britain, in *Linking Networks: The Formation of Common Standards and Visions for Infrastructure Development.* Martin Schiefelbusch and Hans-Liudger Dienel, eds. Farnham: Ashgate, 151–68.

# PART IV
## Conclusions

# Chapter 15
# Key Findings of the Case Studies – A 'Meta-Analysis'

Christian Henrich-Franke and Melissa Gómez[1]

## 1. Introduction

The present volume aims to compare integration processes within different infrastructure sectors throughout the nineteenth and twentieth centuries in Europe. The various contributions develop this comparison using examples of individual infrastructures and their integration. The respective specifications of these examples are identified and explained. Furthermore, similarities and differences are identified on both a general and a specific level.

In this contribution, a comparative analysis of infrastructure integration has been conducted through the assessment of different (1) time periods and (2) infrastructure sectors. Throughout this analytical process, several important aspects were taken into consideration. First, the analysis of the various examples focuses on the structure of integration and the way these organisations transcended national boundaries. Second, an actor's panorama is presented. This panorama explains the diverse levels of relevance of each actor, the diversity and/or concentration of actors throughout each century, and the characteristics of each group (e.g. expert, user, institutional). Furthermore, the processes of standardisation are explained. The types of agreements that motivated these processes are taken into account. The extent to which these were innovative, or merely adaptations of already successful examples, is considered. This first approach to a comparison reveals the main economic, political and other, diverse causes and goals relevant to the various case studies.

---

1 The authors gratefully acknowledge the assistance of Martin Schiefelbusch in preparing this article.

## 2. Comparison across Sectors and Periods

*2.1 Structures of Infrastructure Integration*

*2.1.1 Across sectors*

Various aspects influence the structures of infrastructure integration within different case studies. Therefore, it is important to recognise the difficulty of classifying these into categories. Constant and complex changes, evolution and redefinition of parameters are inherent to each example. Still, some of the criteria met by the various processes can be defined as tendencies that follow a successful example or paradigm. Furthermore, when comparing the experiences of Europe with those of other continents, the results produced are highly interesting. The situations in these different contexts have been influenced by aspects, which differ greatly from one another. At the same time, they demonstrate similar reactions.

The comparison of structures that lead to infrastructure integration within various sectors reveals that rich and diverse paths have been followed. The great number of failed initiatives is also revealed. Global political and economic situations have been very influential in regard to both the promotion and hindrance of integration. This is demonstrated by the rise and, later, relegation of nation-states as main actors.

It seems clear, in the majority of the cases, that the various periods of the global system and their respective political characteristics strongly determined the development of infrastructure integration. In Europe, for example, internationalised visions of the British aviation sector were a manifestation of the contemporary political situation. In the American context, challenges of transportation integration demonstrate the obstacles posed by the 'realm of politics' and the fact that key decisions were made at the very centre of national politics (Seely).[2] Furthermore, the East African example of transport infrastructure integration demonstrates both the initial responses to the strategic interests of the British Empire, as well as how the lack of political interest, focus and investments prevented its further development (Klopp and Makajuma).

Another main topic of debate in almost every sector was which roles should be played by state and private actors in processes of infrastructure integration. Due to the constantly changing character of this process, it would be biased to categorically classify the methods of different infrastructure sectors. It seems evident that each sector experienced a period of fierce protectionism of national markets, throughout which access to foreign investors was restricted and domestic monopolies (e.g. electricity, railway and telecommunication sectors) were promoted. However, these protectionist periods were highly influenced by the political-economic atmosphere of the time. Later, when the need and opportunity for expansion was recognised, this led to open competition.

---

2   Throughout this chapter, the names of authors given in brackets with no further bibliographical information refer to their chapters in this book.

The development of new institutional structures for infrastructure integration often resulted from the experiences of other infrastructure sectors. An example for this is the case of telecommunication and the Deutsch-Österreichischer Telegraphen Verein (DÖTV), which was used as a model across Europe. Similarly, the aviation industry referred to the Wagons-Lits Company as a paradigmatic example. The electricity sector also took inspiration from the railway industry.

*2.1.2 Across periods*

If we exclude the early years of infrastructure network developments, in which integration occurred independently of international organisations, we observe that international organisations were a dominant force in infrastructure integration throughout both periods. Slight differences between the two centuries can be recognised in the complexity of international institutions and the judicial nature of the different organisations.

During the nineteenth century, non-governmental organisations and conferences were predominant. These focused mainly on individual topics, such as the coordination of timetables for international railway transport. Often, they emerged out of bilateral agreements. Non-governmental organisations were the first to be established within the majority of infrastructure sectors. Governments responded by establishing intergovernmental organisations. The activities of the government usually did not prevent processes of non-governmental integration. Rather, they were supplemental.

Throughout the twentieth century, the establishment of non-governmental and governmental institutions occurred in the reverse order. Government organisations emerged first. This then led – directly or indirectly – to the establishment of non-governmental organisations. We have observed this trend within the railway and telecommunication sectors (see the chapters by Schiefelbusch, Henrich-Franke). The establishment of a supranational body, in form of the European Economic Community (EEC), is an outstanding example. Great enthusiasm was transmitted across all industries in order to achieve European integration. Many common bodies or brands were created, such as the TEE in the case of the railway sector (Schiefelbusch). It seems clear that the nineteenth century was characterised by the important role of nongovernmental organisations and private actors (e.g. markets), and the twentieth characterised by the initiatives and intentions of governmental institutions. However, this would be too simplified a statement, as there are some exceptions.

Overall, the network of international institutions throughout the twentieth century was more tightly interwoven. For the first time, institutional linkages between different infrastructure sectors were established. The EEC attempted to formulate a transport policy to encompass the three modes of inland transportation. International relations, independent of legal status, regularly demonstrated trans-national characteristics. Often, they were carried out between national infrastructure administrations. This reduced the importance of governments, foreign ministries and diplomats as interfaces between national and international levels. Finally, it must be emphasised that international institutions developed independently. Within the discussion of the integration of infrastructures, new topics were individually added to the agenda. These were negotiated by an increasing number of committees and organisations. We can, however, observe continuities across both centuries. Many organisations, which were established in the nineteenth century, continued to exist and function throughout the twentieth century.

There are remarkable differences in regard to the geographical scope of infrastructure organisations and the application area of standards. In the nineteenth century, a 'variable geography' (i.e. different compositions of members) applied, in which industrialised Western European states were the nucleus. Then, integration proceeded pragmatically. Actors tried to find simple solutions for satisfying the economic and technical requirements of trans-border communication and transportation. These solutions corresponded to economic necessities. In contrast, the spaces of integration throughout the twentieth century were politically determined. Participation in international organisations and standardisation often resulted from political opportunism. Especially in the 1950s and 1960s, international institutions were established for either capitalist Western Europe or socialist Eastern Europe. Other areas of integration include regional communities (e.g. Nordic countries).

Another aspect, specific to the same time period, is the detrimental effect that the post-war global order had on market-dominating actors. This was the case for British influence on European civil aviation within reconstructed Europe (i.e. composed of nation-states). Later, in the discussion on African transportation infrastructure integration, the focus on national boundaries and markets would again be present. A constant tension existed between the intention to create a pan-African network and the necessity to build independent states, which would focus mainly on national economic and political interests (Klopp and Makajuma).

*2.2 Actors of Infrastructure Integration*

*2.2.1 Across sectors*
The positions and tasks of stakeholder groups of different infrastructures varied quite dramatically. This resulted from the differing characteristics of the infrastructure sectors. User groups, as an active shaping force, made only a relatively small direct impact on the integration of infrastructure. However, their growing demands for interaction were important drivers of integration.

In the processes of infrastructure integration, national political bodies played an important role. Mostly, they were involved through financing, legislating and regulating these processes, such as in the case of the railway and electricity sectors. In some cases, they even became co-owners or owners. In addition, the aviation sector demonstrates how main political parties played a fundamental role, at least from a propositional perspective. They were aware that a World Air Authority or pan-European airline would be effective for the development – and thus control – of European civil aviation. Government subsidies for the development of private networks in the United States and the lack of support of African politicians to railway projects, due to private investments in road projects, demonstrate the fundamental nature of national policies or political decisions. These effectively promoted or hindered plans for infrastructure integration.

In the international political sphere, we can observe that trans-national political institutions were important in emphasising the issue of international infrastructure integration. Examples include the European Economic Community, the European Conference of Ministers of Transport and the United Nations Economic Commission for Europe, and their efforts to draft measures for putting railway integration into practice. In addition, the League of Nations was a pioneer intergovernmental organisation, which attempted to develop electrical integration while national authorities were imposing restrictive legislation.

Furthermore, the key role played by national network operators and their organisations in integration processes is undeniable. The discussion panels on the development of standards and harmonising policies, which were organised by the Union International de Chemins de Fer (UIC), were of great value. These strove for the revitalisation of the railway sector at a point in time at which it was threatened by growth in the road and aviation sectors. Trans-national enterprises, at least in the liberal era, were the main actors in the electricity sector, despite the activities of national authorities. In the telecommunication sector, the Deutsch-Österreichischer Telegraphen Verein, as the first international telecommunication organisation, served as the foundation for eventual integration processes in other infrastructure sectors. Furthermore, PTT-administrations were considerably active in creating, supporting and promoting unions (e.g. International Telegraph Union), committees (e.g. Consultative Committee for International Telephony, International Telegraph – and Radio – Consultative Committee) and conferences (e.g. CEPT). However, their decisions and measures were not always binding for national authorities.

Expert elites also played an important role in infrastructure integration when initiative of public or other actors was lacking. This was the case in the electricity sector, in which engineering organisations strongly stimulated standardisation and lobbied for a European electricity network. The telecommunication sector serves as another example, in which the trans-national technocratic elite promoted active negotiations in regard to operation, administration and tariffs (Henrich-Franke). In the aviation sector, 'think-tanks' and other organisations established a 'high-powered' discussion group, which would come to be one of their most influential actors. In the United States, a group of engineers and technical experts served not only as a promoter of active integration, but also arbitrated many policy aspects (Seely). Furthermore, the Iron Silk Road case study demonstrates how the concept of 'technocratic internationalism' played an important role in pursuing trans-national railway projects, in spite of political divisions (Anastasiadou).

Last, but not least, some crucial actors in contemporary integration processes have recently emerged and guided important projects and visions. In Africa, for example, these include multinational institutions (e.g. African Development Bank,

World Bank) and foreign governments. The government of the People's Republic of China has been increasing both its investments and its influence on policy-making within the African continent.

### 2.2.2 Across periods

The number and diversity of actors increased as the complexity of international infrastructure relations increased. For both periods, we can observe different constellations of private, political or techno-administrative actors. We can also observe that expertise, and experts, played a much more prominent role throughout the twentieth century. Since the interwar period, epistemic communities of techno-administrative actors (i.e. experts) have emerged and shaped the international integration of infrastructures. They often bypassed political authorities and negotiated the integration of infrastructures (e.g. standards) amongst themselves. The telecommunication sector provides another example of the distinct twentieth century phenomenon of expertise. In this sector, a technocratic elite played a vital role in determining infrastructure standards (Henrich-Franke). In the electricity sector, as well, an informal organisation of engineers was in control of in- and outflows (Lagendijk).

The two periods differed with regard to the question of who initiated the integration of infrastructures. In the nineteenth century, this task was usually left to non-governmental actors or members of infrastructure administrations. In the twentieth century, governments, politicians and/or bureaucrats from international organisations promoted the issue. This can be observed within the aviation sector. The Labour Party, with its idea of 'Wings for Peace', called for the establishment of a World Air Authority, which would have a monopoly over primary international routes. In the development of the railway sector, the state played a more relevant role as a stakeholder throughout this second period (Schiefelbusch). Still, politicians often only had influence on the integration of infrastructure for limited periods of time. After initiation, projects were turned over to the experts. Nevertheless, many infrastructure projects required the involvement of political entities, simply because of their magnitude (e.g. financial, geographical). This is shown clearly in the example of the Soviet Union where the central government boosted infrastructure in its vast territory with the aim of becoming a model for a future socialist Europe and Asia (Gestwa). Furthermore, trans-national infrastructures exercised a certain fascination over political actors. Politicians often entered into negotiations on behalf of their own interests, thereby interfering with decision-making processes. The techno-administrative actors preferred these processes of decision-making to be rational (i.e. based on facts and expertise).

It should be emphasised that users of infrastructure were an important motor for their integration. Infrastructure networks would not have been established or expanded were it not for their users, who bought travel tickets or used international communication and transportation services. Organised, trans-national user groups, which attempted to exert influence on infrastructure integration, constitute a phenomenon specific to the twentieth century.

## 2.3 Content of Infrastructure Integration

### 2.3.1 Across sectors

Technical, operational and judicial aspects were the primary issues during both periods. For example, in the case of the electricity sector, the international transmission of electricity was the central concern (Lagendijk). In the aviation sector, a World Air Authority was established in order to operate all civil aircraft by the same standards (Zaidi). Furthermore, in the railway sector, integration was not defined as being only the uniformity of track gauges. Agreements on rolling stocks, energy supplies, operative regulations and signalling were made, among others (Schiefelbusch). Tariffs, however, often proved to be a difficult matter, as neither governments nor administrations wanted to forfeit control over their incomes (Henrich-Franke).

Some, differentiated concepts of integration were not always dichotomous. Sometimes, these were even complementary. As demonstrated by the telecommunication case study, integration was defined as the standardisation of different economic parameters and the institutionalisation of forms of governance in the technical, operative, administrative and tariff fields (Henrich-Franke). Furthermore, within the railway sector, integration represented all processes that reduced friction in social and technical systems. At the same time, mainly in the development of rail passenger services, standards were considered for the establishment of common conditions for international cooperation (Schiefelbusch).

'Fragmentation' was another key concept in the integration of infrastructure. However, it was not always conceived of in the same way. For the electricity sector, it clearly referred to the different technological, political and legislative aspects, which prevented or slowed international exchanges. In the African case study, fragmentation comprises all of the difficulties and complexities of integration. These result from – among others – high costs, geography and diverse, powerful external interests (e.g. interests of former colonial powers, foreign governments).

### 2.3.2 Across periods

The comparison of content is limited in its scope, as the particular problems differed remarkably between the two periods. In the twentieth century, infrastructures were usually at a higher level of development. The interdependency between content of infrastructure integration (e.g. different types of standards) and institutional designs are striking. Tariffs and infrastructure policies, especially, were negotiated solely within intergovernmental and/or supranational institutions. In the late nineteenth century, according to Ambrosius, regulative quality and safety standards were negotiated and enacted by intergovernmental committees, while experts appreciated the benefits of cooperative standardisation in the fields of operation, administration and tariffs.

On a very general level of abstraction, we can observe three prominent differences between the two centuries:

1.  The lucrative mass markets for technical equipment were shielded from foreign competitors by incompatibility standards throughout the twentieth century. Integration and fragmentation were two sides of the same coin. We can observe this in the case of electricity, in which foreign access to domestic resources was restricted in favour of national interests. At the same time, this restricted the international flow of electricity (Lagendijk).
2.  The harmonisation and standardisation of infrastructure policies, or the establishment of common infrastructure policies (e.g. EEC policies in the fields of transport and energy) emerged only during the twentieth century. In general, theoretical conceptions for infrastructure policies were unheard of in the nineteenth century.
3.  Aspects of inter-modality were negotiated primarily during the twentieth century. This increased following the introduction of the container in transport chains in the 1960s.

### 2.4 Reasons and Objectives for Infrastructure Integration

#### 2.4.1 Across sectors

If we compare the reasons for or aims of infrastructure integration processes, we can observe some similarities. It is possible to classify them mostly as political, economic or societal objectives. This does not, however, apply to all factors.

Politics are an element common to almost every case explored in this book. Often, they demonstrate the role played by national governments throughout the period of study. When considering the role of governments in the international standardisation of infrastructures, we must take into account the different understandings of 'the nation' and 'the international' inherent to the nineteenth and the twentieth centuries. Many nations were 'invented' (Hobsbawn) during the nineteenth century by making use of infrastructure systems and their potential to induce national coherence and exercise influence. In light of this, a new understanding of borders emerged. Thereafter, the nation-state was a precisely defined territory. Throughout the nineteenth century, infrastructures were a key means for demarcating the national, as opposed to the non-national or international.

Indeed, in some sectors, priority was given to the development of networks within national boundaries. Examples include the electricity sector, although there was traditionally little resistance in cross-border transmission between European countries, and the railway sector, in which every network focused on a national scale and market. This concept of borders, as developed in the nineteenth century, remained widely accepted throughout the twentieth century. National and international spheres and territories continued to be clearly defined and separated.

Following World War II, it became increasingly obvious that nation-states were no longer capable of meeting all the needs of their populations. In light of the emergence of new actors, such as technocrats and experts, national governments had to react. They could either allow technocrats greater capacity for action or address the issues on a political level.

Generally, increasing user numbers led to the expansion of infrastructures across borders. Intentions to improve relations among European nation-states were often expressed in form of proposals for integration, which was seen as a key element of achieving stability and peace within the continent. This is clearly reflected in the railway vision for a peaceful, civil European empire, or in the motivation to establish an international, durable body within the aviation sector.

Economic reasons and motivations were just as strong as – if not stronger than – political ones. The motivation to increase profits, reduce costs or operate more efficiently was a common and recurrent argument. For the railway industry, it was a way to combat the loss of market shares and regain relevance in the transport sector. Several incentives seem to have promoted cross-border integration in the electricity sector. These included the increase in benefits, augmentation of supply security and the increased transparency of prices, as stated by the European Commission in its European Act. Efficiency, meanwhile, motivated the aviation sector, as European commercial aviation was considered to be too inefficient. In the United States, as well, the goal of increased efficiency motivated transportation integration.

Furthermore, the enthusiasm of new and effective methods for integration spread across most infrastructure sectors. This occurred as a common market and European coordination became increasingly evident. Within the railway sector, for example, important stakeholders harboured scepticism and fears for their commercial independence. In spite of this, many optimistic voices supported European integration as a path towards their own predominance on the transport market. Similarly, within the electricity sector, the Marshall Plan – opposite the European Coal and Steel Community, which also represented a main impulse – would promote European cooperation. It understood electricity as being an important contributor to economic growth. European culture, meanwhile, served as the foundation for the increased standardisation within the telecommunication sector. In this case, cooperative approaches enabled paths for the negotiation of new types of standards.

### 2.4.2 Across periods

To understand similarities and differences in the reasons for and objectives of infrastructure integration, we must consider that the access to or use of infrastructures was an elite phenomenon in the nineteenth century and a mass phenomenon in the twentieth century. Both the number of users and infrastructure services increased dramatically. This had an enormous impact on the quantity and quality of standardisation measures and infrastructure networks. Furthermore, by the twentieth century, infrastructures, their standardised equipment and regulations could look back on a longer history. Operational rules for trans-national networks, which had been established in the nineteenth century, needed to be adjusted to new requirements. The ITU Convention, the basic document for all international telecommunication services, is a good example (Henrich-Franke).

Economic reasons: Economic reasons and objectives were the most important factors for the integration of infrastructures, regardless of period, sector and/or geographical region. However, their relative importance was slightly higher during the nineteenth century. Infrastructures were to open up new international markets, increase the exchange of goods, services and people, and simplify all sorts of international transactions. Infrastructures were necessary for economic integration and for raising average prosperity levels. The expansion of infrastructure networks and the increasing coordination of infrastructure services was largely a response to an increased demand for cross-border distribution. Such was the case of the Polish borders, in which the gradual liberalisation of the border regime occurred after 1956 as a result of the developments in foreign tourism and transit (Komornicki).

During both periods, infrastructures provided a necessary condition for economic integration. However, this alone was not sufficient. In Africa, throughout the early independence period, the perceived need to focus on national interests conflicted with attempts by many countries to promote pan-African integration (Klopp and Makajuma). In twentieth century Europe, economic objectives even resulted in a partial fragmentation of international infrastructures. Many lucrative national markets for telecommunication and transportation equipment were protected against international competition. This was accomplished by making use of incompatibility standards. The networks, however, remained inter-connectable.

Political reasons: Political reasons and objectives differed considerably between the two periods in regard to their type, number and importance. Throughout the nineteenth century, the demonstration of political power was an important objective. This could be accomplished, for example, through the hosting of an international conference. During the twentieth century, political objectives were much more complex and influential. The attempts to establish a global or European political authority (e.g. League of Nations, United Nations, Council of Europe or EEC) had an enormous impact on infrastructures. Such international regimes, which established a general institutional framework, were absent during the nineteenth century. The reduction and construction of boundaries, as well as changes in their nature, were equally important. Furthermore, the case study of the Soviet Union shows how infrastructure was a clear instrument of economic and political integration, which was thought to prepare Soviet people for the transition to communism (Gestwa).

Societal reasons: Societal reasons were of secondary importance. Examples include the expansion of telephone lines for private international calls or the easing of travel restrictions. International actors only implicitly expressed societal interests. However, tourism in the twentieth century is an exception to the rule. Integrated infrastructure services and networks were necessary for people to experience the different cultures and countries of Europe. Such experiences and trans-cultural contacts were important factors in the process of European integration. The general public was not, as we have explored in the chapters of this volume, aware of the negotiations on the integration of infrastructures. On the contrary, international connections were often taken for granted. Trans-national infrastructures were used for cross-border travel and experiences. Processes leading to their genesis, however, were hardly noticed. It seems appropriate, then, that the European Union's infrastructure policies (e.g. trans-national networks) are among the most unknown fields of the EU.

## 3. Conclusion

As we have seen, the comparison across sectors and periods provides interesting insights into the history of infrastructure integration. In spite of all the differences, it was possible to isolate some similarities on an analytical 'meta-level'. Some aspects might even be classified as causal, at least rudimentarily. They include:

- Non-governmental and intergovernmental organisations, including supranational ones, are two sides of the same coin. The establishment of one soon triggers the establishment of the other. Furthermore, both are highly interdependent when negotiating trans-national connections and standards;
- For all infrastructures, we can observe long-term expansions of international structures and standards. This begins with the nucleus of very simple technical or operational standards. Then, processes are activated that subsequently lead to complex duties, such as the development of common infrastructure policies (i.e. spill-over effects);
- Without supranational bodies, national infrastructure markets tend towards protection against international competition once a critical size is reached (i.e. lucrative mass markets);
- Economic objectives are the driving forces behind both the international integration of infrastructures and their fragmentation.

Such causal relationships could serve as a basis for a theory or modus of infrastructure integration. However, much research still remains to be done. It seems advisable to further consider failed projects of integration and/or military infrastructures. The establishment of international military infrastructures, especially, is a highly neglected field of research.

In our reflection of the methodology of this comparison, we must also mention its limitations. The content of infrastructure integration – and, particularly, of operational and technical standards – was hardly comparable. This applies, at least, when comparing details or when examining across periods. Structures, actors, objectives and visions proved to be much more adequate parameters for comparison.

# Chapter 16

# Experiences from Comparative Historical Analysis – Conclusions for Research and Policy

Gerold Ambrosius, Hans-Liudger Dienel and Martin Schiefelbusch

The chapters that make up this volume deal primarily with the establishment of connections among infrastructure networks and services across political boundaries. In the following, the 'integration of infrastructures' or 'infrastructure integration' is defined with that in mind. The aim of this contribution is to present several perspectives for future research on the basis of the preceding chapters and recently published literature on the topic. Its structure follows the analysis of historical events (e.g. the establishment of a railway connection between two countries), in which a differentiation is usually made among the causes, the 'courses' and the effects.

Then, we will consider the new research questions and political implications that result from the conclusions of this book.

## 1. Causes of the Integration of Infrastructures

If, to begin, we assume that decisions on transboundary connections of infrastructures and persons were made consciously, then the causes consist of the motives, intentions and/or aims that political and economic actors followed in promoting these connections. A sectoral and epochal comparison shows that these were fundamentally similar. However, general statements on how much the significance of the individual motives or aims differed among the sectors and/or changed over time are still made on the basis of a relatively narrow empirical basis. The question of whether the financial aspect (e.g. operative gains or losses of individual transboundary connections) played a larger role in the nineteenth century than in the second half of the twentieth century can only be answered with a plausible assumption. This also applies to the issue of whether the cultural aspect, in the sense of international understanding, carried a particularly high priority in the period following World War II relative to the period preceding it. We know that such superordinate motives or aims hardly played a role for the experts that planned and implemented integration projects. Rather, their focus was on overcoming concrete technical problems.

We know little about the relative significance of the various motives and/ or aims to those who initiated and implemented the projects on a political level and/or how they changed over time. Further case studies on specific projects of integration are necessary in order to gain a deeper insight into this issue. Furthermore, an increased understanding of national planning and implementation cultures would accompany greater knowledge about the motives or aims of the relevant actors. Motives and/or aims provide insights into strategic conduct in national and international policy- and decision-making processes. They are an expression of the interests of those that were involved with integration projects. Also related are the reasons for the failure of integration projects, which were initially to be tackled. Such failure can be the result of political structures and/or political processes, but can also result from changing interests.

Do political and economic actors only enforce that which results from the technical and economic rationality of infrastructure networks? It remains to be seen whether such a question can even be answered. In any case, historical research must refer to the theoretical approaches of network economics when dealing with the internal logic of infrastructure networks.

## 2. Courses of the Integration of Infrastructures

The integration of infrastructures, as a political process, occurs within the framework of national and international political structures. New case studies may provide initial insights into national political styles and/or cultures. More comprehensive knowledge on national policy- and decision-making processes in this field of politics would be especially interesting. It was one of the first fields, alongside traditional diplomacy, in which international cooperation was practised.

Infrastructures were and are characterised by hybrid rights of ownership and propriety, in the sense of public and private operation of infrastructure companies and managements. Relatively little is known about the practice of this cooperation. Solid knowledge about the working methods of international regimes and organisations, within the framework of their formal structures (e.g. the Universal Postal Union, International Telegraph Union and World Bank, which played an important role in financing transboundary projects in developing nations after World War II), is available. We also possess well-founded knowledge about the high levels of diplomacy and political cooperation during the nineteenth and twentieth centuries in general. However, we know little about how these international political regimes were interconnected in the concrete political field of infrastructure integration, how these linkages changed over time and which sectoral and regional/continental differences existed. If one considers that, especially in the nineteenth century, the first international connections often were established on a private or semi-governmental basis, government institutions only gradually intervened in the integration process, the cooperation between private and government institutions continued throughout the twentieth century and will probably strengthen further in the foreseeable future, it becomes clear that historical research can make a significant contribution to an increased understanding of this element of international or even global relations.

Thanks to recent studies, first information is available on many aspects of inter- or transnational cooperation in the field of infrastructures. However, as was mentioned previously, comprehensive knowledge is not yet available. Research gaps continue to grow, as soon as regions outside of Europe are taken into consideration. This applies to the cooperation among the countries of these regions, but also to global cooperation. In regard to infrastructures, this touches upon many aspects that are central to research on the history of international politics and relations in general: centre vs. periphery, colonial powers vs. colonies, 'South' vs. 'North', First and Second World, etc.

The aim of historical-empirical research should be, not only to study individual integration projects or to compare them in an intra-/intersectoral and/or interepochal sense, but also to develop a typology of infrastructure integration. The matters (standards) that had to be adjusted for the establishment of a transboundary connection must be placed in relation to the structures and processes occurring on national and international levels. It is a matter of making general statements about which factors, within a specific national and international context, influenced which matters and enforcement forms and how. How can the relationship among structures, processes and standards be characterised? The standards are an expression of political contents that were negotiated. Until now, the development of a historically founded typology has not been accomplished. This is due to the fact that the number of studied integration projects is still relatively small. It is also due to the fact that historical research has focused too little on the models developed by systematic-theoretical standardisation research. The minimal use of such models and theoretical deliberations presents another deficit of historical infrastructure research.

## 3. Effects of the Integration of Infrastructures

In comparison with the causes and courses of the integration of infrastructures, even less is known about the effects thereof. This constitutes the greatest research gap. The main reason for this might be that many effects of integration cannot be precisely defined. The fundamental question, of whether developments would have proceeded similarly without the progressing establishment and expansion of transboundary infrastructure connections, presents itself. It should be grossly differentiated among political, economic and sociocultural or societal effects.

The political course of individual integration projects and the resulting political effects are mixed, since the first connections of the nineteenth century marked the start of an inter- and transnational cooperation that continues until today. This is a matter of more or less continuous processes, the various projects of which were based on the previously gained experiences and simultaneously resulted in new experiences and effects. Even if, in the meantime, knowledge is available on how the cooperation within international organisations and national institutions was rehearsed and made to be routine over time, it still cannot be claimed that the topic has been comprehensively researched.

However, one fundamental problem is that the formal or official policy- and decision-making processes can be reconstructed, albeit incomplete, while the informal or unofficial processes cannot. This problem affects the so-called epistemic community of experts in public administrations and private organisations, which plays an increasingly important role over time. Official correspondence and/or published protocols often did not document that which was discussed in 'private', confidential conversations.

The question of whether different national political styles assimilated over time also goes largely unanswered. After all, the administrations responsible for the infrastructures were permanently in more or less close contact since the outgoing nineteenth century. The obvious question to ask is what the significance of infrastructure integration was for political integration. Did infrastructure integration launch a process that gradually transferred from 'only' technical to increasingly 'political' integration projects? Were forms of cooperation practised here that were later relevant for 'high' politics?

If one considers the huge significance of economic motives or aims to the integration of infrastructures, already in the nineteenth century, the lack of studies on the economic effects becomes apparent. There can be no doubt that faster, safer and cheaper transport of goods and information, persons and capital (due to improved communication) had profound economic effects on local, regional, national or even larger economic areas. However, what these were, in detail, remains largely unknown. Which areas profited and which lost? Where did agglomeration occur and where did deglomeration occur? How did the production structures change? Did an alignment of prices, interest and perhaps even wages occur?

Many more questions of this kind, which have been hardly or insufficiently answered by historical research to date, can be posed. The specific research on infrastructure integration must view itself as being part of general research on economic integration. It should, therefore, refer to the appropriate models (econometric) of modern economic history. In the meantime, we have discovered much about how international institutions and organisations agreed upon necessary standards. However, less is known concretely about project management or the forms of financing of the individual integration projects.

In regard to the sociocultural or societal effects, a broad field of research presents itself, to which we can merely refer to here. Historical research, for example, has provided results about the changing perceptions of space and time and, with that, changing 'mental maps' of increasingly interconnected regions and countries throughout the nineteenth century. In regard to the second half of the twentieth century, research has focused primarily on mass tourism, which became possible with improved transport and communication systems, and the resulting effects on cultural identity. The societal effects of the increasing infrastructural interconnectedness of large areas are multifaceted and offer many opportunities for further research. One must pose the question of where to draw the line. There is a certain risk that the direct correlations between infrastructure integration and general societal effects could remain too vague.

## 4. Taking Infrastructure Research to the Next Level

These reflections demonstrate that, in spite of the progress made in historical research on transboundary infrastructures (including the chapters presented in this book), much still remains to be done. The following reflections aim to outline some important future research fields. In doing so, we hope to make it clear that research in this area is not only of historiographical interest, but also an activity from which today's infrastructure policies could take useful inspiration.

The impression should not arise that historical research on the integration of infrastructures is at square one. There are already many solid results available that should not be shaken by further studies. However, one can still say that infrastructure research, which also focuses on the specific aspect of the connections among infrastructures across political boundaries, has only developed recently. The questions related to the historical integration of infrastructures are therefore astoundingly current.

The previous synthesis described the complexity of historical analyses in this field, which relates to two characteristics of the topic. The first is the number of involved stakeholders and the number of sources to be analysed (cf. the methodological chapter by Christian Henrich-Franke). This is even truer with respect to international projects. Material from several countries, as well as from a growing number of international and supranational bodies must be consulted. All of these stakeholders need to be understood, not only by themselves, but also in their relation to other entities and to the social, political and cultural 'environment' (to use an intentionally broad and unspecific term) in which they operate and with which they interact.

Research in this area, therefore, must consider how to access and efficiently handle such a variety of sources. There are probably no easy and one-for-all solutions. Language skills and the ability to quickly obtain an understanding of a project's political, economic and cultural 'setting' are likely to be of growing importance. It remains to be seen which project formats may be the best to approach such tasks and which methodological standards are required. The growing online availability of archive registers and sometimes even of the sources themselves certainly helps to also do research on topics that are literally further afield. However, in doing so, one must avoid letting the easy availability of material gain too much influence on the selection of sources and/or on the outcome of the analysis. The importance of the last issue is set to increase, as paper-based communication is substituted increasingly by much less tangible ways of oral and digital exchange.

The need for a more intense exchange between historiography and systematic-theoretical research has been discussed above. This is not necessarily an easy task, as the endeavour to identify patterns and typical elements is quite likely to be at odds with the aim to capture each case on its own and in all of its dimensions. However, a certain level of abstraction and condensation is necessary, if the 'message' of such academic work is to have an impact on policymaking. In this sense, a typology as a possible outcome can do much to inspire the present. A typology can serve not as a complete representation of the past, but as a tool for approaching the present and the future differently. It can assist in looking beyond the particular paths and settings of one's personal experience.

Second, the complexity of integration processes increases considerably when the dimensions of 'causes' and 'effects' are considered alongside the 'process' as such. It is here that the greatest knowledge gaps exist. Perhaps this is not surprising, in view of the difficulties in drawing a line between the infrastructure sector, while taking consideration of the 'wider context', as well as the challenges of distinguishing causes and effects. Yet, if we look beyond the processes of harmonisation and coordination occurring in the respective sector, it is clear that infrastructure development is driven by societal needs and creates new opportunities and problems, which in turn lead to new demands and, perhaps, even to new solutions. The well-known policy cycle is also a helpful illustration in this field.

At this stage, we will leave the question of where precisely the boundaries lie between the institutions and processes that provide infrastructures (i.e. 'infrastructure system') and the shaping societal forces. Instead, we shall reflect more upon the multitude of perspectives, actors and processes involved in infrastructure integration that might be 'left out', even though they should find due consideration in future research:

- Globally speaking, infrastructure integration is predominantly framed positively (e.g. easier communication, accessibility, exchange and economic growth). However, research should not be ignorant of the fact that integration does have its drawbacks and dark sides, as is discussed in at least two of the contributions to this book (see Engels and Gestwa). In the more recent past, protests and the concerns of citizens, residents and disadvantaged local authorities accompany almost every infrastructure project. Many readers will be able to confirm this on the basis of their own experience. Without discussing the reasons for this here, it is clear that this dimension should not be left out of analytical research. This includes the retrospective look at projects that lie much further back in time;
- The outcome of infrastructure integration can be seen in terms of international agreements, completed networks and other facilities. However, research should also look into unseen parts of the story, in particular into 'failed' projects and the reasons for their demise;

- The study of integration as a process, when done predominantly by experts working in specialised technical or administrative institutions, risks leaving aside important dimensions in favour of adopting a rationalistic and mechanical understanding of these developments. While such a view is certainly inspired by reality, it does not explain everything. Many infrastructure projects require the involvement of political entities simply because of their size (e.g. financial, geographical). Additionally, they exert a certain power of fascination on political actors, which may be classified as one of the 'dark sides' or just as normal human behaviour. How can these disturbances to otherwise 'rational' expert-based processes be captured by retrospective research? Considering the issues of power and accountability, however, it may well be that retrospective research delivers more trustworthy results than contemporary analysis;

- In connection, it may be worthwhile to reflect upon project size as a factor in itself. This is particularly relevant to comparative approaches. In this respect, political appeal and feasibility should be considered as distinct influences, which can support, but also contradict one another.

Ultimately, these questions are still posed today in regard to the establishment or expansion of trans-European networks, regional transboundary connections in Europe, the badly needed connections among countries on other continents and even the global organisation of air and radio traffic.

When looking at the present and into the future, we observe additional levels of complexity emerging. First, infrastructure projects are increasingly subject to rising demands, conflicting expectations, planning and implementation delays and financial problems. While public memory often 'forgets' such issues after some time, research must take a closer look at such difficulties. Last, but not least, this can contribute to an understanding of how they might be avoided in the future. Doing so, however, requires detailed technical knowledge, the presence of which is not guaranteed in many research layouts.

Second, the matters discussed, in this book, as national, bilateral or European matters are, in reality, global topics. How does infrastructure integration in Africa (Klopp and Makujama), Asia (Anastasiadou and Tympas, Gestwa) or North America (Seely) differ from infrastructure integration in Europe? Do we, in light of the globalisation of trade and communication, need a more 'global' view and/or institutions with this perspective? Finally, what can past experiences from specific fields, such as railways and roads, tell us for the development of infrastructure for the virtual world?

# Index

For Product Safety Concerns and Information please contact our EU
representative GPSR@taylorandfrancis.com
Taylor & Francis Verlag GmbH, Kaufingerstraße 24, 80331 München, Germany

www.ingramcontent.com/pod-product-compliance
Ingram Content Group UK Ltd.
Pitfield, Milton Keynes, MK11 3LW, UK
UKHW021619240425
457818UK00018B/645